Kwaito Bodies

Xavier Livermon

Kwaito
Bodies

Remastering Space
and Subjectivity
in Post-Apartheid
South Africa

Duke University Press *Durham and London* 2020

© 2020 DUKE UNIVERSITY PRESS
All rights reserved

Designed by Courtney Leigh Baker
Typeset in Whitman and Canela by Westchester Publishing Services

Library of Congress Cataloging-in-Publication Data

Names: Livermon, Xavier, [date].
Title: Kwaito bodies : remastering space and subjectivity in post-apartheid
   South Africa / Xavier Livermon.
Description: Durham : Duke University Press, 2020. | Revision of the author's thesis
   (doctoral)—University of California, Berkeley, 2006. | Includes bibliographical
   references and index.
Identifiers: LCCN 2019033523 (print) | LCCN 2019033524 (ebook) | ISBN 9781478005797
   (hardcover) | ISBN 9781478006633 (paperback) | ISBN 9781478007357 (ebook)
Subjects: LCSH: Kwaito (Music)—Social aspects—South Africa. | Popular music—Social
   aspects—South Africa. | Urban youth—South Africa. | Post-apartheid era—South
   Africa. | Human body—Political aspects—South Africa. | Sex role—South Africa. |
   Queer theory—South Africa.
Classification: LCC ML3917.S62 L58 2020 (print) | LCC ML 3917.S62 (ebook) |
   DDC 781.630968—dc23
LC record available at https://lccn.loc.gov/2019033523
LC ebook record available at https://lccn.loc.gov/2019033524

Cover art: *Kwaito Culture*. © Neo Ntsoma Productions. Courtesy of the artist.

To my dad, Eugene Robertson

Contents

Acknowledgments   ix

Introduction. Waar Was Jy? Yeoville circa 1996   1

1. Afrodiasporic Space
Refiguring Africa in Diaspora Analytics   29

2. Jozi Nights
The Post-Apartheid City, Encounter, and Mobility   57

3. "Si-Ghetto Fabulous"
Self-Fashioning, Consumption, and Pleasure in Kwaito   92

4. The Kwaito Feminine
Lebo Mathosa as a "Dangerous Woman"   122

5. The Black Masculine in Kwaito
Mandoza and the Limits of Hypermasculine Performance   155

6. Mafikizolo and Youth Day Parties
(Melancholic) Conviviality and the Queering of Utopian Memory   188

Coda. Kwaito Futures, Remastered Freedoms   224

Notes   235
Glossary   239
References   243
Index   259

Acknowledgments

Completing a book is a monumental task that, while solitary, is never accomplished alone. I would like to take this opportunity to thank the many people and organizations that were a part of making this study possible.

This study would not have been possible without the financial support of numerous funding organizations. I would like to thank the Graduate Division at the University of California, Berkeley, for the Chancellor's Opportunity Pre-doctoral Fellowship and the Summer Humanities Grant. Summer and Year-long FLAS (Foreign Language and Area Studies) Fellowships in Zulu were central in helping me to acquire important language skills. I was also fortunate to receive funding in the form of Rocca predissertation and dissertation fellowships through the African Studies Center at the University of California, Berkeley. Fulbright funded this study twice. I received a Fulbright-Hays Summer Group Project Abroad Fellowship to study Zulu, as well as a Fulbright Institute for International Education fellowship for dissertation research. Completion of my dissertation was greatly aided by the Dissertation Fellowship in the

Department of Black Studies at the University of California, Santa Barbara. Once I completed the dissertation, the Carolina Postdoctoral Fellowship for Faculty Diversity gave me time to begin revising the manuscript for publication. While at Wayne State University, this project was generously supported with two grants from the Humanities Center, where I served as a Faculty Fellow. Further support was provided by the National Council for Black Studies Gender Research Grant and the Washington University Law School Center for the Interdisciplinary Study of Work and Social Capital, where I served as a research scholar for the Black Sexual Economies Project. While at the University of Texas at Austin, my work was supported by the Center for Women's and Gender Studies Faculty Development Program, the College of Liberal Arts Summer Research Grant, and the Humanities Center Faculty Fellowship.

This study is the product of numerous intellectual and personal engagements with a variety of scholars. First, I would like to thank my dissertation committee, Jocelyne Guilbault, Charles Henry, Percy Hintzen, and Trinh T. Minh-ha, for their mentorship and support over the years. While in South Africa, I enjoyed the intellectual engagement and support of Christopher Ballantine, Angela Impey, and Keyan Tomaselli at the University of KwaZulu-Natal; and David Coplan, Achille Mbembe, Danai Mupotsa, and Sarah Nuttall at Witwatersrand University. Gcobani Qambela of the University of Johannesburg and Zethu Matebeni at the University of the Western Cape have also been key interlocutors throughout the years. Ideas that would become central to this study were developed in courses taken with Gina Dent, Utz McKnight, Patricia Penn Hilden, and Françoise Verges. I would also like to thank my undergraduate professors Bennetta Jules-Rosette and George Lipsitz, who first sparked in me the idea that my intellectual curiosity could translate into an academic career. The African Studies Center at UC Berkeley was a constant source of support and intellectual engagement, and I could not have imagined my graduate experience without Martha Saavedra. While at UC Santa Barbara, I was fortunate to receive constant feedback on early drafts of these chapters from Neda Atanasoski, Ingrid Banks, Peter Bloom, Emily Cheng, Douglas Daniels, Melissa Forbis, Gaye Theresa Johnson, Claudine Michel, Stephan Miescher, Sylvester Ogbechie, Roberto Strongman, Daphne Taylor-Garcia, and the late Clyde Woods. From my time at the University of North Carolina, Chapel Hill, I would like to thank Renee Alexander Craft, Laura Halperin, Ken Hillis, Ashley Lucas, and JoAnna Poblete. At Wayne State University, I was fortunate to be surrounded by a dedicated and engaged group of scholars, including Lisa Alexander, Melba Boyd, Sarika Chandra, Robert Diaz, David Goldberg, Eboe Hutchful, Ollie Johnson, and Lisa Ze Winters. While at the University of Texas at Austin, this work has been intellectually and emotionally supported by a number of wonderful colleagues, including Simone Browne, Ann Cvetkovich, Tshepo Masango Chery, Lyndon Gill, Ted Gordon, Kali Gross, Sue Heinzelman, Neville Hoad, Omi Osun Jones, Cherise Smith, Pauline Strong, Lisa B. Thompson, and Hershini Bhana Young. I would also be remiss if I did not thank the collective group of scholars who make up Black Performance Theory and my colleagues from the Black Sexual Economies project at Washington University in St. Louis.

I have been fortunate to work with a number of graduate students who have also supported this project, including Ali Neff from my time at UNC Chapel Hill, William Banks from my time at Wayne State University, and Auzimuth Jackson and William H. Mosley from my time at the University of Texas at Austin. Each of them provided feedback, editorial assistance, and moral support throughout various stages of this project.

There is no way that this book would have been possible without the support and labor of particular friends and intellectual interlocutors. Words alone cannot describe the depth of gratitude that I feel toward these people. I am truly blessed to have them in my life. The first group is my graduate cohort Libby Lewis and Ivy Mills. The second, my South African crew, Mpho Mokoena, Neo Mothlala, and Amos "Sello" Mutloane. Lastly, there are the people who took the time out of their busy schedules to read and give extensive feedback on revisions to this manuscript, and their contributions are noted and appreciated here. Marlon M. Bailey, Mireille Miller-Young, Matt Richardson, and Omise'eke Tinsley all provided mentorship and feedback through numerous rounds of revision. I would also like to thank the anonymous reviewers for pushing me to make this manuscript the best it could be. And I thank my editor, Elizabeth Ault, for being a strong champion of this project from its inception.

Lastly, friends, family, and community have sustained me through this project. Thanks to Tricia Wilson; my godson, Zay Wilson; my mom, Carol Robertson; my sister, Tameka Tataw; my nephew, Eugene Tataw; and my brother-in-law, Elvis Tataw. To my sprawling and loving family in/from Virginia—the Langhornes, the Livermons, the Robertsons—thank you for all of your support and love over the years. For my partner, James, who supported me through the most difficult moments of the revision process, I am eternally grateful. I also want to acknowledge the people of South Africa, the San Francisco Bay area, and Austin who nurtured me intellectually and emotionally throughout this long process. Unfortunately, my father, Eugene Robertson, did not live to see this manuscript completed. While he was alive he provided immeasurable levels of emotional care and support. I dedicate this book to him. I know he would have been proud to read it.

Introduction

Waar Was Jy?

YEOVILLE CIRCA 1996

Yeoville is a neighborhood that sings to the heart. From the moment of my first sojourn in Johannesburg, I found myself drawn to the neighborhood's vibrant, resonant streets. Beyond the front door of my student guest house right off of Raleigh/Rockey Street, I began to explore the neighborhood with gusto. I found a city so cosmopolitan, so alive: the Congolese barbers who gave the best haircuts, the café that served the fluffiest pastry, the bar with the best—and cheapest—drinks, the clubs with the most eclectic mixes. So much seemed possible at night in Yeoville; I wanted to be a part of that possibility.

Twenty years later, those nighttime memories bring a sly smile to my face. I remember hearing the 1996 Skeem hit "Waar Was Jy?" (Where were you?) for the first time in a taxi from Johannesburg's main train station as I made my way back from Maseru to Yeoville. The unmistakable thumping of the music, that familiar four-on-the-floor house signature, the electric keyboard melody, and the voice of Ismael, the region's ruling new voice in the heady, rhythm-driven soundscape that is *kwaito* dance music, soaring. Ismael seemed to taunt us listeners with his question: *Where were you?*

> Waar was jy?
> Where were you?
> One o le kae?
> Bowukuphi? (Skeem 2013)

Songs like Skeem's breakthrough hit provide resounding demonstrations of kwaito's play with various sound "traditions"—in this case, the pop and bubblegum era of 1980s township disco—to root itself in a long genealogy of South African music. Now Skeem and early kwaito itself, particularly songs like "Waar Was Jy?," are undoubtedly considered part of those traditions. Listening to the song in our present moment reveals how music shifts and establishes new orders of time. The song, produced in the 1990s, implored its listeners to remember where they stood in the 1980s; today it has become self-referential: "Where were you when you first heard this song?" It has become the index for the very question it asks its listeners to consider, over and again, in its four-minute window of time.

Yeoville by night is an analytical entrance into the kinds of possibilities I witnessed in the bodies of Black post-apartheid youth when approached through kwaito as a time-folding cultural form. These performance practices, set in time to kwaito's beats, are critical to this study. A night out in the Yeoville of the 1990s was an ephemeral experience, the conditions for which will not be duplicated again. Dressed and ready to see what the night had in store, I did what many young Black South Africans living in the newly integrating inner city of Johannesburg did at the time. My crew of young cosmopolitans and I wandered from club to club, absorbing the vibrancy of the streets among other young Black people who may have lived in the townships, or may have been university students, or may have been young professionals seeking new thrills in the otherwise familiar landscape of urban nightlife. We were all looking for opportunities to perform new freedoms.

If any one spot in the social landscape was representative of this early burgeoning Black youth culture and its pulsing nightlife, it was House of Tandoor. Located in the heart of Yeoville, this two-story nightspot was the place to see and be seen. It was decorated in what can only be described as Rastafarian Afro-bohemian chic: a prominent red-yellow-green paint palette complemented by various portraits of Bob Marley covering the walls. Downstairs, kwaito roared from the speakers and the appreciative crowd swayed, grinded, and pulsed to the latest and a few classic kwaito hits. But the fun of House of

Tandoor was always in the blend of the upstairs and downstairs parties. Upstairs was an open patio, rooftop-like deck where one could watch the crowds on Rockey Street stream by while dancing to the latest dancehall, hip-hop, and R&B hits.

While radio (first Bop, then Metro FM, and later YFM) may have been the first medium for the latest sounds that would make up the mélange of post-apartheid Black urban culture, nightspots like Tandoor were often spots to perform and experience these songs. The mix of people roaming from place to place in the area ensured a particular kind of openness to the new, the different, and the experimental. I remember the shock and thrill of seeing openly queer African men and women occupying public space on Rockey Street, a site that seemed somehow open to performances of queer sexuality without being marked as queer space. As I explored my own budding sexuality—enjoying the prospect of flirtatious eye contact and lingering stares, releasing into a hand placed flirtatiously on my shoulder or arms gripped tight around my waist as a new partner and I danced in unison—I embraced a sense of possibility that seemed unimaginable in the spaces I had left behind. In those more familiar corners of my world, queer space was racialized as white and heavily segregated from nonqueer space. This book considers how kwaito artists, and their performances, fans, and representations, illuminate variegated discourses and paradigms of freedom for Black South Africans in a post-apartheid world.

Based on the richness of a long-term, immersive ethnographic engagement, I argue that kwaito—an important site for investigating the politics of contemporary South Africa—is an alternative technology that mediates the Black body. To be clear, my investigation of kwaito is not meant to argue that popular cultural formations overturn the neoliberal contract of post-apartheid South Africa. Nor do I suggest that kwaito's constellation of performances necessarily operate in a one-to-one relationship with liberatory political economic frameworks. Instead, kwaito is important because of the inherent limits of national recognition for Black South Africans, incorporated into the "new" South Africa through a political compromise that fosters and perpetuates continued classed, gendered, sexualized, and racialized inequality. As a cultural formation, kwaito is a means through which young Black South Africans negotiate the limits of national recognition.

Kwaito serves as an alternative site that recognizes the political dimension of age practices (particularly the performative practices) of Black South African youth, even when its work is not meant to directly challenge global capital or the post-apartheid state. It is important to revisit and reevaluate the importance of kwaito's popular performance practices, and the power of Black urban post-apartheid popular cultures as a whole. While, to the outside listener, kwaito may seem apolitical, unengaged, reactionary, or even regressive, the music and its culture stand as an important site of politics whose outcomes are far from guaranteed and fail to easily fit into a resistance/co-optation binary. For Black urban youth, the cultural practices that arise in kwaito are as critical a site for political practice as any political and economic institution or action.

This study is the culmination of more than fifteen years of engaged, carefully contextualized ethnographic research into kwaito performance. As an African American, I immediately recognized that there is much about kwaito that is familiar. The sounds, fashion, and dances are closely related to a combination of contemporary, globally circulating Black musical styles and performances. Yet there is also something distinct about kwaito; it is not simply an imitation of international Black musical and cultural styles. My first engagements with kwaito were primarily as a fan. In the mid-1990s, I took numerous trips to Johannesburg to participate in the fledgling kwaito club scene. I also attended house parties and concerts and collected a substantial archive of early kwaito cassettes. Early on, I could not comprehend the lyrics, but my body understood the music. I danced to it in clubs and at house parties. I cheered for it at live concerts. There is something indescribable in its pace, and the atmosphere created in kwaito party spaces freed me from the strictures of masculinist posturing that had previously dominated my approach to popular and social dance. As I danced I performed a kind of freedom in concert with the other kwaito bodies. Kwaito's rhythms, even and seductively slow, produced body movements in me that were lithe and sensual, with emphasis on the sway of the disarticulated upper body and the pelvis. Simultaneously, kwaito songs seemed to require movements in my lower body that at times appeared to outpace the music itself. This combination of sensuality and freneticism produced in me a joy while listening and dancing, giving me permission to perform my body differently, to deliberately release toughness and "cool" and instead celebrate abandonment and freedom from any cares. In this space I and the other kwaito bodies present could inhabit our subjectivities differently. These early experiences taught me that kwaito is more than simply music. Like other Black musical practices, kwaito is a culture that facilitates the creation of something excitingly new and politically disruptive for post-apartheid South Africa. Together, these performances form the lens through which I analyze kwaito.

Shortly after the first democratic elections in South Africa, media and cultural critics began to recognize the controversial impact of kwaito music. Kwaito arrived on the scene just as South Africa was undergoing its transformation from apartheid. Within the nation-state, sociocultural developments—

which in the past might have been confined to township space—were, like Black people themselves, actively reshaping the South African landscape. And yet kwaito has become more than just a musical genre created by young Black South Africans in the wake of apartheid's collapse. These cultural practices are a fertile site for contemporary South African youth agency; their strength becomes an index by which we can better understand Black life. At the same time, kwaito has become symbolic of the crisis, both real and imagined, of youth culture in post-apartheid South Africa. Issues of violence, sexuality, urbanization, and political economy (especially the rise of neoliberal socioeconomic policies) are just some of the crucial realities of concern for post-apartheid youth. Young Black South Africans use kwaito to negotiate the future of their nation. The tensions in public discourses surrounding kwaito evidence the political nature of the culture as it resists and reimagines the post-apartheid state, particularly in relation to the shifting class, racial, ethnic, and gendered politics of various Black communities. For these youth—as well as a host of associated communities who create and consume the culture—kwaito serves as a site of cultural politics that brokers the constant interrogation and remaking of Black subjectivities.

Both for contemporary African Diaspora communities and for the world at large, South Africa has emerged as a nation full of significant symbolic capital. The racial inequalities of the apartheid era marked South Africa as the last vestige of a system of racial colonialism remarkably similar to slavery for global Black communities. Apartheid South Africa served as a stark reminder that people of African descent everywhere were not truly free. The work of South Africans to heal the wounds of its racial past and to reconcile its stark economic inequalities is of interest to global Black communities as they seek to reconcile less stark, if not less troubling, forms of inequality in their own countries.

Kwaito practitioners, fans and artists alike, often describe kwaito as a soundtrack to freedom both personal and political. As Ishmael (of Skeem) remarked, "Kwaito was an expression of liberation. It was a freedom of some sort. People were becoming more and more themselves. It was a take it or leave it attitude. This is the way I am, the way—ngingakhona" (Nkosi 2014). I asked one of my close friends, Tumelo, to elaborate on what his hopes and dreams were in the 1990s. What did the end of apartheid mean for him? "For me, I just want to have a nice life. I want to give my family a nice life. I want my own house. I want my mom's house to be comfortable. I want my younger brother and sister to have opportunities I did not. I want to live in a nicer place. I don't want to struggle financially" (personal communication, 2004). I looked around at his current circumstances—he lived at home with his parents in a working-class area of his township, there were clearly things around the home that needed fixing, and he struggled with un(der)employment. While there was a level of security in the home (there was always food, electricity, water), it was clear that ten years into South Africa's transition, these hopes and dreams of his remained stalled. I asked him if freedom was only or primarily about improvements in material comfort. "No," he remarked. "I am also a gay man. I am also Black. I am also Tswana. So being able to be openly gay, especially here in the township, is an important part of my ability to be free." Here, Tumelo indexes two critical themes to the way Black youth would describe freedom: through questions of remaking space and subjectivity.

In order to understand the cultural labor that kwaito participates in, it is necessary to sketch some specific aspects of South Africa's apartheid and post-apartheid history. During the 1980s, political and economic tensions were at an all-time high in South Africa. The reinvigoration of the anti-apartheid movement from within the country (oftentimes forgotten in the shadow of the resistance from outside) made the costs of perpetuating apartheid increasingly untenable to the National Party government. A South African friend, exiled to Southern California in the wake of the upheaval, framed the South African freedom struggle through the inevitability of violence: "We really did not believe that political negotiation would bring an end to apartheid; we all assumed that South Africa was headed towards a war, a long bloody war" (personal communication, 2006). While this study focuses on the cultural dimensions, rather than the immense historical specificity, of the negotiated compromise of South African politics, the basic premise held that political power would be shared equally (the mantra of one vote, one person), while efforts would be made to create a more equitable economic structure post-apartheid. During the near revolutionary conditions of 1992 and 1993, the African National Congress (ANC) was tasked with demobilizing urban protests in order to ensure that the negotiated settlement would proceed (Bond and Mottiar 2013).

The South African situation both resonated and clashed with those of the majority of African states that had negotiated and fought for independence in the post–World War II period. Like much of postcolonial Africa, leaders had to make efforts to encourage the entirety of the populace to buy into the idea of the new nation, and to encourage a racially, ethnically, religiously, and linguistically (to name just a few of the major societal cleavages of South Africa) diverse set of people to identify with the new South African nation. Unlike much of postcolonial Africa, however, South Africa had a relatively advanced middle-income economy that was highly integrated into the world economy.

In addition, South Africa had to negotiate the transition of this economy under the simultaneous triumph of global capitalism (symbolized by the collapse of the Soviet Union) and the rise of neoliberal economic reforms (symbolized by the numerous structural adjustment programs imposed on developing economies).

To address the idea of nation building, the ANC and the South African media adopted the notion of the "Rainbow Nation." Desmond Tutu first articulated the concept in a speech against apartheid state violence related to Cape Town city elections in 1989; it was a call to recognize what Tutu called the "Technicolor" of South Africa (Tutu 1994). Rooted in political theology, the concept spoke to the interconnectedness and diversity of South Africa's people, and the folly of the suggestion that any one group of people could singlehandedly determine South Africa's future: "Remember the rainbow in the Bible is the sign of peace. The rainbow is the sign of prosperity and justice, and we can have it [peace, prosperity, and justice] when all the people of God, the rainbow people of God, work together" (Tutu 1994, v). Tutu would elaborate on the notion of the "rainbow people of God" and coin the term "Rainbow Nation" after South Africa's first general elections. Nelson Mandela would lend the construction further credibility when he used it to describe South Africa during his first months in office. Critics of the phrase, and its adoption as a particular form of governmentality post-apartheid, emphasize the ways in which it has spiraled into a facile and apolitical celebration of multiculturalism, imposed from the top down, that rarely poses a challenge to entrenched social divisions (Habib 1997; Valji 2003).

To address the idea of economic redistribution, the ANC, along with its partners the South African Communist Party and the Congress of South African Trade Unions, pursued what it called the Reconstruction and Development Programme (RDP). The RDP was an attempt to address the fundamental socio-economic inequalities inherited by the ANC government. Central to this strategy was the coordination of social services meant to alleviate poverty. The goals of the program were housing provision, electrification, clean drinking water access, healthcare, land redistribution, and a massive public works campaign. RDP policy documents suggest that early on, however, such a program could be sustained only by steady and increasing macroeconomic growth. Hence, the RDP is often viewed as an attempt to negotiate the tension between the need to provide greater social services outlays while entering into a global economy dominated by neoliberal orthodoxy. That this policy was abandoned in 1996 in favor of Growth Employment and Redistribution has been heavily critiqued in progressive political circles as an abandonment of the ANC's (and to a larger extent the anti-apartheid movement's) redistributive economic goals. The adoption of a more orthodox neoliberal strategy is subject to a number of interpretations, which broadly fall into two camps: the first suggests a fundamental shift toward a "pragmatic" ANC strategy to address post–Cold War global realities, and the second suggests that the ANC government capitulated to an ideologically driven economic strategy that could not help it meet its development goals. These critics feared that the policy would serve primarily to enrich the Black political and economic elite, a danger Frantz Fanon identifies as "the pitfalls of national consciousness" (1963). Perhaps the most obvious example of this pitfall is the Black Economic Empowerment scheme, which promised to diversify the complexion of capital in South Africa, but has, in fact, benefited a small elite to the detriment of more systemic economic redistribution.

In South Africa, the adoption of neoliberalism began in earnest with the late apartheid regime, where hegemonic internal struggles in the National Party gave rise to the emergence of a strong technocratic wing. This group ultimately collaborated with the interests of the English-speaking economic elite and global capital. As part of the negotiation process, the ANC dropped promises to nationalize banks, mines, and other forms of global capital; agreed to pay back $25 billion of apartheid-era inherited foreign debt; ensured the independence of the central bank; joined the General Agreement on Trades and Tariffs; and pursued an $850 million loan from the International Monetary Fund, replete with the usual conditions. As Bond and Mottiar (2013) explain, after the ANC came into power, "privatization began in earnest; financial liberalization took the form of relaxed exchange controls, and interest rates were raised to a record high. . . . [T]he ANC government granted permission to South Africa's biggest companies . . . to move their listings to London. Where corporate profits were retained in the country, they did not feed into investment in plant, equipment, and factories. Instead the financialized economy encouraged asset speculation in real estate and the Johannesburg Stock Exchange" (286).

Taken together, the critique of the "Rainbow Nation" and neoliberal macroeconomic policy speak similarly to the dissatisfaction with the incomplete nature of South Africa's political and economic transition. Scholars across traditions and trajectories (Bond 2000; Desai 2002; Hart 2002; Kunnie 2000) agree that the influence of the elite on South Africa's transition from apartheid led to the reproduction of structural inequality: the result rendered entire classes of Black South Africans disposable and lacking agency to make political change. While some material changes have unfolded in post-apartheid South Africa (despite the critique that only the color of the leaders has changed, not their values), it is clear that the process of transition has inadequately addressed the needs of the majority. Even though the South African economy has grown in size (from a GDP of $57 billion in 1985 to a forecast GDP of $510 billion in 2015), it has actually become more unequal in the post-apartheid period, with the Gini coefficient increasing from .59 in 1994 (CIA 2013; World Bank n.d.) to .65 in 2005 (CIA 2013), .67 in 2006 (World Bank n.d.), or .70 in 2008 (Institute for Justice and Reconciliation 2012), depending on which study is cited (see Harmse 2014).

Perhaps the most contentious issue, which illustrated the increasing inequality of South Africa's transition and demonstrated the ANC's fixation on neoliberal orthodoxy, was the handling of the Marikana mine workers' strike by the South African security forces. In the post-apartheid period, platinum mining (of which South Africa has nearly 90 percent of the world's known reserves) has outstripped gold and diamond mining to become the main source of mineral wealth contribution to the country's GDP. Workers based in Marikana, a platinum-rich mine belt northwest of Johannesburg, had been engaged in a tense wage strike for several days when, on August 16, 2012, South African security forces fired on them, killing thirty-four and wounding at least seventy-eight (the exact number of wounded is unknown). It stands as the single most lethal use of force by the South African Police Service against civilians since the Sharpeville massacre of 1960. Initial reports of the killings circulated by police spokespeople (and subsequently amplified by the corporate-dominated media) suggested that the massacre was a justified police action. They claimed that the miners were destabilizing the post-apartheid economy; that their strike was "illegal" given the miners' lack of support by their own union; and that the miners had attacked security forces in a drug-induced haze inspired by a traditional healer who had given them magic medicine that would ostensibly ward off bullets. The police and press recirculated representations of (rural) Black men, who make up the majority of the mine's workforce, as irrational and violent (long a trope of the apartheid regime's propaganda) as justification for this particular police action. However, under further scrutiny, the story woven by the security forces, Lonmin (the British-based owners of the mine), and high-ranking members of the South African government began to fall apart. Evidence emerged that many of the dead miners were shot in the back and at close range while fleeing. The miners, increasingly frustrated by the cozy relationship between their traditional representatives the National Union of Mineworkers, the ANC government, and mine management, made a logical decision to shift their allegiance elsewhere.

The event laid bare the legacies of elitism that haunted the new ANC government. Far from a crazed crowd charging irrationally at police, the strikers assembled peacefully until they were surrounded and corralled by state security forces. The fact emerged that careful coordination and communication between the mine owners, state police forces, and important representatives of the ANC government (including current South African president Cyril Ramaphosa) had led to the massacre. This was not an impromptu response to a fast-changing and uncertain situation. Rather, the massacre was a result of the pressure Lonmin placed on the South African government to manage worker unrest, and to ensure that profits continued unabated. Instead of acting as an impartial force or protecting the rights of workers to strike, the ANC government colluded with the forces of global capital to end the strike at all costs. The deaths of thirty-four (and the wounding of countless others) were simply a small price to pay in order to keep the status quo of corporate profits and its ancillary distribution to the new political elite intact.

Marikana paints a very grim picture of what South Africa's post-apartheid transition really means; it is symptomatic of the failures of the post-apartheid state. The constitutional promise to shed apartheid's past seems inconsequential in the face of an economy that remains shockingly unequal. Scholars should examine the possibilities inherent in the post-apartheid transition, however limited or constrained they may be by an entrenched political economic hierarchy. Popular culture is a critical space for these alternative political practices. By looking at the urban Black youth politics of post-apartheid South Africa through the lens of kwaito, we excavate the cultural formation as an important site of struggle. To do so evidences the agency of kwaito practitioners located in the culture: the power to contest and negotiate their post-apartheid conditions.

A cultural study of kwaito illuminates numerous dimensions of the political work it does for its fans and practitioners. In particular, kwaito facilitates alternative performances for Black bodies as it contests the constraints of geography, race, class, gender, and sexuality as they are produced through South Africa's post-apartheid transition. For example, Arthur Mafokate's song "Kaffir" (1995) caused controversy precisely because it trafficked in the racially derogatory term for Black South Africans. Many older Black South Africans felt the word should disappear, not be reanimated to create a more assertive Black subjectivity, one that demands, as Mafokate does in the song, that white South Africans no longer call him by the racial slur. Much of early kwaito was about an assertive (re)claiming of space, an announcement of arrival, a demand to be listened to, to not be discounted because of youth or circumstance

(Nkosi 2014). Angered by a lack of nominations at the 1995 South African Music Awards and sensing that this was a discounting of eruptive forms of youth culture, Mafokate simulated anal sex with his dancers on stage (McCloy 2006). The song itself and its irreverent performance of it by Mafokate and his women dancers serve as one example of how kwaito created the space for multiple forms of contestation. Kwaito offers a vision of what freedom will look like from the perspective of young Black South Africans; more specifically, it tells us how this freedom will be performed and what practices will sustain and nurture this freedom.

Kwaito Bodies: Remastering Space and Subjectivity in Post-Apartheid South Africa is an examination of the figure of the youthful Black body in post-apartheid South Africa's public culture, using kwaito—its representation and the sites of its consumption—as its point of focus. While kwaito serves as an organizing theme for the text, it goes beyond an analysis that frames kwaito solely as a music genre; instead, it probes kwaito as a cultural performance that is widely and sometimes contradictorily interpreted and mobilized. Three key contributions animate the text. First, *Kwaito Bodies* emphasizes new consumption practices over the (dis)continuation of struggle politics. Hence, an important contribution is the reorientation of consumption within South Africa's urban youth culture as itself a politic. The analysis hinges on these youths' sense of having the right to consume: to move through the city, to self-fashion differently, to take pleasure in being able to party, to work with the past in seemingly irreverent ways, among other interventions. A second contribution is the borrowing of theoretical and analytical thinking from African Diaspora studies for the South African context. Kwaito is a music that is as local as it is Afrodiasporic. Hence, the text brings questions as inflected in the scholarship emerging from the Caribbean, Black Britain, and the United States to bear on the contemporary South African context. Discussions of Blackness as it relates to gender and sexuality against the backdrop of colonial racism elsewhere help us recognize the ways South Africans are managing the post-apartheid legacy of entrenched heteropatriarchy and racialized inequality. The effect is to move away from the general tendency to theorize South Africa from a position of exceptionalism. Lastly, the analyses emerging from the fields of popular music studies and race/gender/sexuality studies commingle here to animate this text. Hence, *Kwaito Bodies* is the first post-apartheid study of self-fashioning and the mediation of the body and pleasure that arises from the practices of popular performance and social dance. This intersectional analysis prizes the nuanced and contradictory relationship between popular performance as an embodied practice and vectors of power mutually constituted through race, class, gender, and sexuality. In particular, I investigate the ways gender and sexuality are constituted and the multiple yet contradictory ways that Black people—including Black queers—put kwaito to work.

Remastery

To consider what these processes of freedom might entail, I borrow the verb "remaster" from musical production, yet I use it in two different senses. First, I use the verb to speak to subtle shifts in the political economy and the meaning of those subtle shifts—like the process of music remastering, the term is used here to refer to "a much more subtle approach as opposed to remixing since mastering is very limited, you can only do so much. Indeed, remixing involves manipulating separate tracks to create altered vocals and instrumentation. . . . [R]emastering involves taking an original source and cleaning up any sound imperfections, not adding new elements" (O'Toole 2009). It speaks to the racial colonial implications of the post-apartheid transition, the difference that may not be so different after all (Hall 1993). In what ways must freedom in South Africa return to the racial colonial project (the original form—the master), to a process of remaking racialized colonialism? In this sense, I speak of freedom remastered as a performative technology deployed by kwaito bodies to signify differently in a context that is not so different after all. Secondly, I use "remaster" by literally breaking it down into its component parts. What might it mean for freedom to be remastered, to have a new master, for it to signify something new within the residual context of racial colonialism? Significantly, what do these performances of freedom look like when they are remastered by kwaito bodies? At stake is an acknowledgment that the performance of freedom is always contested and that different sociopolitical interests emerge to contest its meaning and implementation in contemporary South Africa.

Is it possible to remaster freedom, to recuperate something in the way we listen for it, to perform it differently, to understand that aural innovation begs to be enacted? Or does the act of remastering reproduce the residues of colonial capitalist hierarchy, offering something that sounds like freedom but is in fact just a clearer copy of an unsatisfying original? This study considers what kinds of future playlists for South Africa have been bequeathed through kwaito and the constellation of performances it has inspired. What has kwaito enabled in the public sphere and how have contemporary South Africans extended its legacies? To reiterate, my use of remastery as an organizing metaphor relies on multiple echoes in the term's meaning. Borrowing from the practice of sound production, I first consider what it might mean to literally enhance freedom in the studio that is post-apartheid South Africa writ large. Second, what might it mean for freedom to have a new master, for the visions of freedom for South Africa to be orchestrated by different bodies? If we consider that placing freedom in the hands of young Black South Africans would also mean its remastery in some significant ways, then we recognize these definitions as symbiotic. In other words, to site freedom in kwaito bodies requires a remastering of freedom, its enhanced reformulation.

While kwaito bodies often insisted that their goal was not political, it is clear that their performances enacted a new political vocabulary: one that was often at odds with various dominant post-apartheid political vocabularies. Writing about the Kenyan postcolony, Keguro Macharia (2016) suggests that what the nation needs is the development and nurturing of alternative political vernaculars. For Macharia, political vernaculars "announce a conversation about politics. They are words and phrases that assemble something experienced as the political and gather different groups around something marked as political. . . . [T]hey create possibilities for different ways of coming together . . . and they also impede how we form ourselves as we-formations, across the past present, the future." Frustrated by the delimited vision of political vernaculars, Macharia calls for new imaginations of freedom and love grounded in practice for contemporary Kenya. Most importantly for him, Kenya needs to develop "political vernaculars untethered to state imaginaries."

Most compelling in Macharia's essay is his call for an *ethics* of freedom and love *grounded in practice* that looks beyond the state as a monopolizing interlocutor. One of the limitations of post-apartheid South Africa has been the substitution of the white heteropatriarchy–led apartheid state with the Black heteropatriarchy–led post-apartheid state. A further limitation has been a political vernacular that cites agency in the heteropatriarchal ruling elite, while constructing citizens of South Africa as little more than recipients of state beneficence. Here, political vernaculars are chained to the state. But what might it mean to tether a post-apartheid political vernacular to kwaito bodies? How might we heed Macharia's call to consider alternative political vernaculars situated in cultural formations, embedded in embodied practices, and committed to an ethic of freedom?

The concept of remastering allows me to imagine the alternative political vernacular that Macharia calls for. I combine the music technological concept of mastering (used as an idea of enhancement) with the Afrodiasporic understanding of mastering (used as the hierarchal relationship of colonialism). In mastering an art (mastery as skills or craft)—here performing on the dance floor and sartorially—kwaito artists and fans remaster (as in rework) heteronormative ideals in the kwaito world. In doing so, kwaito bodies open up the potential to subvert the master-servant hierarchies, the violent history of which produce contemporary South Africa.

Much of what is being remastered through kwaito bodies are ideals of subjectivity and space with perhaps a hope for shifts in materiality. S'bu Nxumalo (known professionally as "the General"), YFM's first music manager and founding editor of *Y Magazine*, suggests that the material hopes of kwaito bodies were misguided: "We didn't understand it [freedom]. We equated this new struggle to bling but we didn't ask why. Why the bling? Why the gold? What does it mean?" (Nkosi 2014). However, kwaito bodies did understand that freedom was at least partially rooted in materiality. The terms of that materiality were decidedly not disruptive to the functioning of capital, but Black youths' demand for material comfort through the existing capitalist system revealed an awareness of economic inequality and a desire to shift that inequality in some way. Hence, they hoped, perhaps in hindsight naively, to remaster the signs and symbols of wealth to work for kwaito bodies. Perhaps more importantly, kwaito was a space for remastering subjectivity. When asked to describe what kwaito meant to Black South African youth, Zola, a top-selling kwaito artist, had this to say: "kwaito for me defines something different. For one, we were trying to figure out who we are and where do we fit in this role of the democratic South Africa as young people" (NPR 2006). Hence, kwaito is a space to remaster forms of subjectivity for Black youth. This study is concerned with the implications of remastery for Black queer youth, but kwaito created possibilities for remastering a number of Black subjectivities related to class, ethnicity, race, nation, gender, and sexuality. Lastly, kwaito is a cultural formation that serves the processes of remastering space. As Nxumalo states, "Kwaito [is a] rebel voice, man. And it's not so much about what you're saying, it's the doing. It's about freeing up spaces" (Nkosi 2014). This study is invested in the doing—the performative as a way to consider how kwaito has been critically important in remastering space, in this case, the club, the neighborhood, the airwaves, in order to create freedom for Black youth.

Kwaito in the Scholarly Imaginary

This book makes three explicit contributions to the scholarship on kwaito. These include (1) a redefining of the concept of politics in kwaito and Black popular culture; (2) a retheorization of the diaspora paradigm as an epistemic practice for understanding kwaito's ontology; and (3) an introduction to the theory of "kwaito bodies" as a concept that goes beyond racialized and gendered subjectivities to elucidate how the bodies of kwaito performers and consumers are sites of subversion, visuality, and sensuality that frame the performative notion of being a Black youth. While journalistic accounts of kwaito emerged in the mid-1990s, scholarly work on kwaito did not begin to appear until the early 2000s. Significantly, early work on kwaito often engaged either implicitly or explicitly the longer genealogy of scholarly work in South Africa that focused on popular music, (Black) youth studies, urbanization, and political studies. For the purposes of my examination, it is useful to identity three themes that emerged in much of this work as a way to recognize both the present study's debts to this earlier work and its critical points of departure. Each of these points concerns the need to identify the political dimensions of kwaito as a cultural practice.

First, much of the early work on kwaito identified, explained, and critiqued the ways kwaito culture embraced consumer culture (L. Allen 2004; Bosch 2006; Coplan 2005; Peterson 2003; Pietilä 2013; Santos 2013; Steingo 2005, 2007; Stephens 2000; Swartz 2008). To the extent that kwaito could be linked to eager consumerism, the literature dismissed the musical form and its adherents as lacking the critical consciousness needed to address the challenges of post-apartheid South Africa. They evaluated the politics of kwaito as either nonexistent or retrograde, claiming that kwaito failed to enact the resistance to (post-)apartheid politics that shaped earlier popular musical cultures. Furthermore, they expressed an implicit, if not outright explicit, panic, convinced that the music trafficked in a kind of capitalist consumption that reinforced rather than challenged the neoliberal state, and revealed a seeming false consciousness among Black youth. Through this lens, kwaito closely mirrored the state's shift to neoliberalism and was political only in its drive to reproduce these relations of inequality (Steingo 2007). To the first generation of kwaito scholars, much of the flamboyant materialism of kwaito represented a form of "bootstrap capitalism" (Peterson 2003, 210) as well as the "internalization of late capitalism and neoliberalism, the disappearance of the political (in the conventional sense) and the colonization of people's consciousness" (Steingo 2007, 33).

Although many of these arguments provide important nuance for some of the larger analyses of neoliberalism, capitalism, and consumption—especially in relation to Black youth cultures—many of these critiques are nevertheless premised on a desire for a more easily recognizable social consciousness and popular progressive politics in Black youth-driven popular cultures. While lamenting the presence of the market in kwaito, other scholars (Coplan 2005; Peterson 2003; Swartz 2008) located redemptive politics in the very textures of capitalist consumption. As Bhekizizwe Peterson suggests, "For one the pre-occupation with consumption can be interpreted as an acceptance of the larger societal ethos that informs many South Africans of different backgrounds and ages. Alternatively, even in its most nihilistic forms, the celebration of consumption in Black youth may attest to their courage and commitment not to give in to the conditions of poverty and strife found in the townships" (2003, 210). Much of this scholarship worked against the tendency to read consumption as social deviance or acquiescence, even at its "nihilistic" extremes, instead championing Black youths' refusal to be reduced to conditions of poverty. The key to much of this work was the "language of aspiration" (Nuttall 2004, 439) that argued that materialism and consumption in kwaito should be read as a reclamation project (Coplan 2005; Swartz 2008). Black youth were reclaiming the spoils of apartheid for themselves even if such displays were politically "diluted" (Swartz 2008, 25) and "self-aggrandizing" (Coplan 2005, 18).

More recent work (Livermon 2015; Pietilä 2013; Santos 2013; Steingo 2016) has approached consumptive practices in kwaito with a marked departure from the earlier moralist framework. Central to these studies is the idea that consumption in and of itself is not deleterious behavior requiring reprimand. Drawing from Manthia Diawara (1998), Pietilä (2013) argues that rather than representing simply another ill of "lost" post-apartheid Black youth, the kwaito market is a key site of struggle among today's Black youth for liberation and advancement. Importantly, she suggests that such forms of consumption in relation to kwaito share corollaries not only with contemporaneous Afrodiasporic popular cultures but also with historical South African musical forms. My own examination of consumptive practices and politics in kwaito draws on Diawara's (1998) and Pietilä's (2013) insights to suggest that market participation is key for Black people precisely because such unfettered participation was denied in the African colonial and South African apartheid context. Similarly, my position owes a debt to the theoretical insights of queer geographer Natalie Oswin (2005), who argues that "commodification is therefore neither above politics nor a signification of their end. It is rather a site in which the political is played out, and in a more complex fashion than the supposition of a resistance/capitulation binary permits us to understand" (583). Consumption is configured not only from outside this restrictive resistance/capitulation binary but also from outside of the moralizing conceptual binary that Oswin names castigation/romanticization. In turn, consumption, materialism, and commodification are approached not as processes to overcome but instead as a "productive social force" (583).

While early scholarship on the growth of kwaito culture is very concerned with the intimate dynamic between the local and global, diverse scholarship has examined how shifting global political economies and available new technologies allowed kwaito to emerge at the precise moment of South Africa's political transition (L. Allen 2004; Bosch 2006; Coplan 2005; Hansen 2006; Magubane 2003; Mhlambi 2004; Niaah 2008, 2009; Peterson 2003; Pietilä 2013; Santos 2013; Steingo 2008b, 2016; Stephens 2000). This scholarship is premised on revealing the multiple global dialogues of Black South African youth and evidencing the significance of these global dialogues within local contexts. Much of this early work presents itself as vindication for kwaito, arguing implicitly or explicitly for the "South Africanness" of the musical form and its connection with South Africa's polyvalent pasts. The specter of cultural imperialism, particularly the dominant American strain, is thus key in these debates. Whereas the discussion of the general political import of kwaito often straddles the resistance/co-optation binary, just as pressing for this scholarship is another key binary in the local-global discursive debate: authenticity versus imitation. To the extent that kwaito could be linked to the local, it represented a form of creative resistance to the forces of globalization, and thus was worthy of being admitted to the lexicon of South African popular music. As Thokozani Mhlambi (2004) asks, "Can kwaito—a genre that is largely influenced by certain kinds of music from the United States of America—be considered a distinctly South African musical genre, or is it just part of a mass expansion of a world youth music genre, cloaked in South African forms? Can kwaito be deemed an authentic South African phenomenon?" (116). Most of these studies take for granted that kwaito emerges as a result of globalization.

A subset of these studies, however, argues for the specificity of the type of "global imaginations" accessed and produced through kwaito (Erlmann 1999; Steingo 2016). These scholars do not deny the import of globalization to the rise of kwaito, but they suggest that the globalization of Black youth cultures and South African popular music occurs through processes of diasporic identification (Hansen 2006; Magubane 2003; Niaah 2008, 2009; Pietilä 2013; Santos 2013). In making these claims, they are arguing for different ways to conceptualize the forms of global exchange occurring in Black South African youth cultures more generally, and within kwaito specifically. Thomas Blom Hansen (2006) uses diaspora theory to suggest that kwaito, particularly among Indian South Africans in Durban, becomes a way to assert an Afro-Indian identity. He argues that kwaito reappropriates the Afrodiasporic sounds and styles that had once been removed from its lived contexts and transformed into a desirable global cultural commodity. Tuulukki Pietilä (2013) discusses the historical and contemporary politics of diasporic exchange inherent in kwaito, while Sonjah Stanley Niaah (2008, 2009) emphasizes how these common cultural genealogies, in particular between Jamaican dancehall and kwaito, reveal shared transnational space rooted in the histories and politics of the African Diaspora. Engaging diaspora theory and its importance to the analysis of kwaito rests on these authors' insights. Different strains of globalization take root through multiplicity across the landscape of the African Diaspora. As Zine Magubane states, "The fact that many of the aesthetic practices the purists decry as 'Western' are in fact African American [and more generally Afrodiasporic] in origin" has seldom been considered for the complexity this adds to analyses of popular culture focused on authenticity, cultural imperialism, and purity (2003, 298). Afrodiasporic connections are approached through notions of shared genealogies, political fates, and identifications that create contemporaneous space. In this configuration, Africa is not figured as a static site of origin from which the diaspora delineates and reconfigures Africanness, but as a place of continuous circulation, mobility, and re-making. As Richard Iton (2008) argues, "This approach to diaspora compels us to resist conceptual templates and metaphors that subsidize thinking in terms of seeds and stems, roots and routes, origins and elsewhere, and that promote the problematic reification and detemporalization of 'Africa'" (200). Iton's interpretation of diaspora as "anaformative," resisting "hierarchy, hegemony, and administration" and producing "an alternative culture of location and identification to the state, [along with] dissident maps and geographies" is central to my approach in this study (200).

A final theme that emerges in early kwaito scholarship is that of gender and sexuality within the music's culture (Blose 2012; Coplan 2005; Impey 2001; Mhlambi 2004; Peterson 2003; Ratele 2003; Stephens 2000). Words such as "violent," "misogynistic," and "lewd" often frame the discussion of gender and sexuality in kwaito. Much like the discussion of consumption, gender is often treated in a totalizing binary frame in which "gender" becomes a metonym for "women." This mode reifies and takes for granted the binary construction of "woman" and "man." In this mode, women are presented either as victims of a patriarchal, misogynistic popular culture or as complicit with their own subjection. Meanwhile, the processes of gender formation that mark normative masculinity in kwaito are often left unexamined. Queers are not mentioned at all and thus are rendered invisible as consumers and producers of kwaito.

Angela Impey (2001) provides one of the few analyses of kwaito that argues for women's use of kwaito space as a forum for feminine agency, a sonic space in which to re-imagine gender relations. Peterson (2003) provides one of the few analyses of masculine gender presentation in kwaito culture, situating the music genre as a response to structural forms of inequality that produce masculinities constantly negotiating the desire to transcend given material conditions. On numerous occasions, while conducting this research, I witnessed women and queers (and queer women) enjoying kwaito as performers and fans. Likewise, I witnessed complicated and nuanced engagements with "manhood" and "masculinity." How, then, do I account for these presences within a popular musical form and that is assumed to be violently misogynistic and, by extension, queer-phobic?

In this study, intersectional Black feminist and Black queer theory provides a more subtle and refined critique of how gender and sexuality work in kwaito and buttress two critical arguments. Drawing from various theorists—included among them Audre Lorde (1984), Kimberlé Crenshaw (1989), Patricia Hill Collins (1990), Kamala Kempadoo (2004), and Dwight McBride (1998)—Black youth, who are the primary drivers of kwaito culture and thus of this text as well, are analyzed not only as racial(ized) subjects but also as subjects constituted by gender, sexuality, class, and a host of other modes of difference in a complex post-apartheid milieu. In fact, gender and sexuality often serve as modalities through which race and class are produced, and thus are thoroughly entangled in kwaito as a popular cultural practice. Second, I do not assume the covalence of kwaito and Black cisgendered male heterosexuality; rather, the constructions of various types of genders and sexualities within kwaito are probed to reveal the spaces where forms of heteronormativity are both reified and destabilized. Here, the political lies in the work Black youth do through their embodied practices of race, gender, and sexuality in the realm of (public) kwaito performance.

However, if there is any scholarly text that *Kwaito Bodies* is most in conversation with, it would be Gavin Steingo's *Kwaito's Promise: Music and the Aesthetics of Freedom in South Africa* (2016). In this critical ethnomusicological study, Steingo argues for the political import of kwaito through a reevaluation of theories of aesthetic practices. Steingo and I share an interest in probing the paradoxes that the rise of kwaito culture presents to South Africa. While I attend to the spectacular event and consider kwaito bodies via performance, Steingo attends to musical processes and backyard sociality via technology and circulation. While Steingo turns to Jacques Rancière and the idea of the platform, I turn to intersectional Black feminism and Black queer theory, Afro-diasporic Space(s), and the idea of remastery. Ultimately, both studies contemplate the relationship between aesthetics (or style and cultural practice) and politics. Steingo sets the stage by articulating the vexed quality of the relationship between consumption, political agency, and kwaito art. Distinctively, this study foregrounds African Diaspora theory, gender, and sexuality to open up kwaito as a source of Black pleasure as a politic against the backdrop of years of resistance politics.

Kwaito Cultural Formation in Context

There is much disagreement in the scholarly and popular literature about when kwaito emerged, who decided it should be named kwaito, and the meaning of the name of the genre. These debates have been covered extensively by scholars elsewhere (see especially Steingo 2008a, 2016). Yet it is important to situate the emergence of kwaito culture within the shifting sociopolitical context of post-apartheid South Africa. To understand that context, it is critical to outline the shifts that were happening both on the national scale (the increasing liberalization of late apartheid life) and in the context of shifting global political economies.

If we understand neoliberalism to be a form of sociopolitical governance that did not emerge magically with the post-apartheid regime, but instead was an important feature of late apartheid and the transition period between Nelson Mandela's release in 1990 and the first elections in 1994, then we can recognize these socioeconomic shifts as fundamental to creating the conditions in which kwaito emerges. As Steingo suggests (2008a, 2016), it might be more appropriate to see the emergence of kwaito contextually not simply as the logical extension of political liberation but also as a consequence of shifts in the global political economy and the effects those shifts ultimately had on global popular culture: "I would like to argue that, in historicizing kwaito, it is imperative to consider larger global flows and shifts. In this vein, I suggest that the triumph of neoliberalism and the end of the cold war in the late 1980s were more significant events (or series of events) in the history of kwaito than the end of apartheid" (2008a, 80). Steingo's observation requires us to situate kwaito's rise in relation to the shifts in the global economy, though it might be useful to think of the processes of global economic shifts and internal political compromise as symbiotic, making it difficult to pinpoint which series of events was more significant.

Two processes are of particular concern in considering the rise of kwaito. The first is the urbanization of late apartheid South Africa. Rural areas had long been neglected by the apartheid regime. Attempts to spur industrial development in homeland areas (in order to discourage urban migration and ideologically support the notion of separate development) were never extensive enough to absorb the surplus labor market in the rural areas. In many cases, rural people had little choice but to seek employment opportunities in urban areas; survival in rural areas, mostly impossible, was viable only at the most rudimentary levels. In addition, people fled the rural areas for a variety of other reasons, including political violence and freedom from family strictures.

Urbanization rates in South Africa began to rise exponentially in the 1980s, even before apartheid pass laws were scrapped. This was likely due to the situation of entrenched poverty in rural areas. As petty apartheid laws were relaxed or ignored from the mid-1980s on, South Africa experienced a significant increase in urban migration, exploding the populations of urban and peri-urban areas alike. It is estimated that in 1986–1987 South Africa shifted from a predominantly rural to a predominantly urban population, and that the rate of urbanization reached its peak (3.3 percent) in 1993; today it is estimated to be at about 1.5 percent per year. People were likely to move to township areas either because it was affordable or because they had social connections to support their initial transition to city life. The relaxation of petty apartheid accelerated the change in the composition of many of the urban areas of post-apartheid cities (Johannesburg in particular), where inner-city areas and peri-urban suburbs became spaces for relocating Black South Africans (as well as foreign Black nationals from elsewhere on the continent). These inner-city townships became incubators for new Black South African social identities and cultural practices. Inadequate investment combined with lack of viable job opportunities in these new Black spaces created an atmosphere where creative solutions became crucial for survival and prosperity in and around cities like Johannesburg.

The second process of note in considering the rise of kwaito is the late apartheid regime's adoption of neoliberal technocratic policies, which opened up the national market in unprecedented ways to global products. State disinvestment and privatization ensued; as the public sector retrenched, the private sector was unable to provide sufficiently for the growing numbers of people arriving in urban spaces looking for work. For many Black South Africans, the irony of political freedom and increased mobility within their country meant that they were compelled to move to urban areas for work that was increasingly constrained by the twin forces of deindustrialization and neoliberalism.

The accessibility of global products opened up the increasing availability of both imported music and the equipment and technologies that would feed new movements in South African music. Musical production now required less-expensive production equipment; it became democratized (see Impey 2001; NPR 2006). Local DJs were spinning this global music in the burgeoning club scene in Johannesburg and its townships. The inner-city areas—newly available for residential or leisurely occupation—were important incubators for new cultural formations such as kwaito. Part of this was facilitated initially by the now-defunct station Radio Bop, which broadcasted from the independent homeland of Bophuthatswana using AM signals to reach Black urban communities in Gauteng. Radio Bop curated local and international hits that spoke to urban Black sensibilities and reflected new musical cultures emerging in the late apartheid era. To compete with Radio Bop, the South African Broadcasting Corporation launched Radio Metro, the first apartheid-regime music industry recognition that Black urban populations not only were an unavoidable local presence but also were keenly listening to and engaging with a variety of global musical trends. Because much of this music was imported, and hence not widely available due to prohibitive costs or lack of sufficient product, a thriving economy of bootleg cassettes emerged to satisfy consumer need (see Steingo 2016). The convergence of street entrepreneurship, hyperconsumption, proliferating club and party spaces, increasing radio mediation, and amenable government policy served to satisfy the tastes of urban Black youth in late apartheid South Africa; this convergence set the stage for the explosion of kwaito culture as an exponentially popular mid-1990s phenomenon.

    The progenitors of kwaito were a mix of club DJs, promoters, dancers, and singers—who may have initially honed their skills in other genres (notably township disco and jazz)—as well as music producers and entrepreneurs looking to take advantage of this fledgling genre. In many cases, these tastemakers had been deeply involved in a series of niches within the Johannesburg nightlife scene before converging in kwaito. Early kwaito was infused with the late-1980s bootleg aesthetic; just as entrepreneurs had sold cassette mixtapes of international music on the streets, early kwaito musicians took their experimental mixes to the street corners, the warehouses, and the trunks of their cars. The transportation of Black South Africans in minibus taxis was an important form of urban entrepreneurship that began to experience tremendous growth at this time. Minibus taxis were literally a driving force behind the explosion and growth of what came to be known as kwaito. Responsible for the transportation of millions of Black South Africans throughout the country, taxi drivers became crucial cultural arbiters, mobile DJs whose music selections began to determine what was popular and cutting-edge. Lastly, governmental policy requiring more local music content on radio and television opened a space on the air waves that kwaito was well positioned to fill, in large part because it appealed to youth in ways that other forms of local music could not.

Kwaito and the Vulnerability of Mixed Methods

How do I, a queer African American man, approach the study of culture and society in post-apartheid South Africa? In what ways does my embodied experience as a researcher affect both the kind of data that I might gather and my interpretation of that data? Living and working for decades in South(ern) Africa, I find myself constantly wrestling with the contradictions and privileges of being a queer African American in various spaces, simultaneously embodying diasporic boundary disputes while also coming upon moments in which I am called on to transcend and remake those very boundaries. Speaking of her own research in Nigeria as a "Pan-Africanist African-American Ethnographer," Omi Osun Joni Jones (1996) argues vociferously for a scholarly practice that acknowledges the centrality of multiple forms of dislocation that occur through and within scholarly practice, particularly fieldwork. In order to account for these forms of dislocation, Jones suggests that researchers acknowledge both the embodied nature of scholarly practice and the necessity of "self-theorizing" (137).

Central to my methodological practice of intersectionality are Black feminist, Black queer, and performance studies frameworks as guiding lenses that allow me to keep a close eye on my positionality as a scholar. If we are to take seriously Jones's (2016) assertion that diasporas are often embodied, then performance ethnography becomes a key methodological intervention for Black diaspora studies and thus a guiding method for this study. Taking embodiment seriously—as Black feminism and Black queer theory contend we must—means recognizing and experiencing ethnography as a necessarily sensual or even erotic experience. Telling my story, revealing myself in the text, not only is critical reflexive practice, but also holds me accountable to my political investments in the work as well as to the limitations of my fieldwork. This is particularly vital once we understand that fieldwork is a political exercise that creates the space for advocacy for specific people, ideas, and positions (O. Jones 2006, 343). "If people are genuinely interested in understanding culture, they must put aspects of that culture on and into their bodies" (J. Jones 2002, 7). Dance, nightlife exploration, and media consumption have allowed me to enter kwaito culture, putting it on and into my body. In this way, I become a "co-performative witness" (Conquergood 2002, 351), acknowledging the fact that I am a co-creator—though hardly ignorant of severe power hierarchies in my relationships—of the very practices that I seek to critique and analyze. As I danced at parties and in clubs, purchased cassettes and CDs, and shared in the latest news about various kwaito musicians, I was not just a keen observer.

Instead, I was—and still am—an invested participant, influenced by and influencing how the very communities I participated in critiqued and transformed kwaito culture. Despite my foreignness and queerness, I was often implicitly or even explicitly implicated in these debates. As Dwight Conquergood (2006) argues, I "spoke to and with" rather than "about" kwaito. Hence, this study is a performance ethnography, not in the sense that I predominantly did my ethnographic research among kwaito musicians and performers (although that was an element of the work). Instead, if we take seriously the performance of everyday life, then this performance ethnography is an attempt to understand the ways in which the everyday cultural practices surrounding kwaito and the methods necessary to map those practices are highly performative.

Performance provides the means through which I can explore critical cultural studies. Following Jones's lead, I understand that "cultural studies seeks to reveal the political ideologies wrapped around everyday human behavior and cultural production, giving particular attention to the way race, gender, sexuality, geography and class shape our understandings of behavior and culture" (2006, 342). And critical cultural studies expand this revelation through the theoretical insights garnered from Black feminist and Black queer studies—especially the centrality of intersectionality as a guiding framework—alongside the keen attention of performance studies. These strategically mixed methods provide the tools with which to investigate the space that kwaito creates for performing freedom in South Africa.

In this study, I frame my analysis of kwaito not through an analysis of the musical form but rather through my understanding of kwaito as a cultural formation that provides space for a number of complex performance practices to cohere—performance practices that remaster freedom. This is not to slight the important work of analyzing the musical form; instead it reflects my commitment to understanding how Black youth in South Africa put kwaito to use. Thus, how kwaito bodies (mostly fans but also in some instances musical performers) perform in the space(s) of freedom articulated to kwaito is my concern here. For me, this choice is political, mostly because, in my analysis, fans' enjoyment of kwaito has been an unexamined area of research in academic literature. Focusing on the performance of kwaito bodies allows me to counter reductionist understandings of kwaito style and practice. It allows me to engage how the concept of freedom can be refracted through these bodies, how freedom as an idea can be variously looked at, listened to, and considered through the spaces of the dance floor. I center kwaito fans, especially Black queer South Africans, as leading the charge through their leisure practices toward a more diverse and equitable South Africa. Because these fans have often been overlooked in both popular and scholarly discussions of kwaito, centering their practices engages new forms of knowledge production that simultaneously call for less traditional evidences and methodologies.

Structure of this Book

In chapter 1, "Afrodiasporic Space: Refiguring Africa in Diaspora Analytics," I use the controversy created by Boom Shaka's 1998 performance of the South African liberation song "Nkosi Sikelel' iAfrika" to introduce the conceptual tools central to the study, most importantly "Afrodiasporic Space" and "kwaito bodies." The debates about their performance center on the polarizing question of whether Boom Shaka was making the song fresh for a new generation or was desecrating the memory of struggle by performing it with a kwaito beat. While "Nkosi" is hardly a song exclusive to the South African nation, it has become one of the potent symbols of the New South Africa if for no other reason than its adoption as the post-apartheid national anthem. I consider what it means to perform the nation to a kwaito beat. I use "kwaito" in this instance as a sound metaphor for the voices, dreams, and aspirations of the Y generation (those coming of age in the immediate post-apartheid period). Ultimately, I am concerned with how the nationalist project of freedom might be remastered by those who find kwaito meaningful to their lives.

In chapter 2, "Jozi Nights: The Post-Apartheid City, Encounter, and Mobility," I explore micro-practices of freedom, taking the reader through two nights of partying with my friends in Johannesburg. This chapter reveals how kwaito bodies remaster space through nightlife and leisure practices. The productive possibility of remastered urban space is explored through the concepts of conviviality and encounter that are introduced as frameworks for understanding urban space. Specifically, kwaito helps to create an alternative geography of the city that reveals the nascent attempts of Black youth to remaster the hierarchies of post-apartheid South Africa. In considering how young Black people repurpose and reorganize space, I show how the city becomes both a physical and a conceptual space where inequalities are destabilized by mobile kwaito bodies.

In chapter 3, "'Si-Ghetto Fabulous': Self-Fashioning, Consumption, and Pleasure in Kwaito," consumption in relation to kwaito performance is examined through kwaito festivals, the opening of the Maponya Mall in Soweto, and the performance practices of *izikhothane*. I explore how kwaito bodies remaster freedom through their practices of consumption and pleasure. Given the history of sumptuary laws regulating how Black South Africans could consume, practices of consumption have always been critically linked to notions of freedom. Excess is explored here as a queering of consumption, a way for kwaito bodies to navigate both their marginal material realities and the forms of social censure and moral panic that surround their consumptive practices. Consumption is revealed as a practice that is deeply connected to individual and communal processes of self-creation. What emerges is a politics of exuberance whose very excess escapes the politics of defeat.

In chapter 4, "The Kwaito Feminine: Lebo Mathosa as a 'Dangerous Woman,'" I engage with the performances of Lebo Mathosa, the deceased former member of Boom Shaka who later enjoyed a successful solo career. She remastered freedom through a disidentification with the "bad girl" persona and the creation of an alternative persona of the "dangerous woman." Mathosa is a dangerous woman because she challenged popular conventions about appropriate femininity through both her onstage performances and her offstage persona. I combine an analysis of her concert and club performances, a reading of her music video for the song "Awudede/Dangerous," and numerous interviews with Mathosa in South African media to show how she performed a type of femininity that contravened accepted notions of Black women's performance. In a context of normalized sexualized violence against Black women, her intervention is dangerous because it boldly confronts the risks and possibilities of reimagining Black women's bodies.

Chapter 5, "The Black Masculine in Kwaito: Mandoza and the Limits of Hypermasculine Performance," builds on the analysis of feminine performance to explore the potential of the masculine kwaito body. Mandoza remasters freedom through his disidentificatory performance of the *tsotsi* (thug). As a figure, the thug has been called upon throughout South African history to do particular kinds of work often centered on the amelioration of (post-)apartheid anxieties regarding poor and working-class urban Black men. Focusing on the thug's post-apartheid meanings for Black South African communities, I explore how Mandoza manipulates and uses the commodified image toward performances of freedom that ultimately encounter limitations. I conclude with an examination of how vernacular thug performances might provide spaces for remastered Black queer subjectivities.

Chapter 6, "Mafikizolo and Youth Day Parties: (Melancholic) Conviviality and the Queering of Utopian Memory," returns to the concerns mentioned in chapter 1: namely, the question of what it means to perform the nation to a kwaito beat. To do so, I listen specifically for resonances of the past in contemporary kwaito performance. The performances of the kwaito group Mafikizolo are used to explore the 1955–1963 destruction of Sophiatown, and contemporary Youth

Day parties are used to examine the June 16, 1976, Soweto student uprising. Kwaito bodies perform and queer the past through their ecstatic (and some might say irreverent) performances of memory. In particular, the performances central to June 16 parties consistently renew the ideals of the liberation movement and provide a space for the kwaito body to live out the possibilities fought for by those who rebelled on June 16. That this is misread by the political elite reveals that we must listen carefully and take seriously the voices, dances, and politics of the Y generation's kwaito bodies.

Afrodiasporic Space

REFIGURING AFRICA IN DIASPORA ANALYTICS

Kwaito music exists in, creates, and helps to define what I refer to as "Afrodiasporic Space." An examination of the genealogy of Afrodiasporic exchange culturally, musically, politically, and performatively between Black South Africans and Blacks from elsewhere reveals that the complex interplay and cross-fertilizations of styles within kwaito cannot be dismissed as something new, or as something representing the unfortunate Westernization of contemporary South African youth culture (see Ballantine 1999; Coplan 2008; Erlmann 1999; Hamm 1988; Nixon 1994; Pietilä 2013; Titlestad 2004). Speaking about a 2006 concert given by Jay-Z in Johannesburg, Tsitsi Jaji reveals that the excitement that accompanied the event "looks less like a recent phenomenon of commercial hip-hop's publicity machine (which it may well be) and more like an extension of a long, multidirectional cultural current linking the two nations" (2014, 247).

The concept of Afrodiasporic Space draws from two key traditions within diaspora studies. The first is represented by African Diaspora theorists who conceptualize diaspora from the perspective of African experiences on the continent. Using W. E. B. Du Bois's text *The Negro* (2001) as a point of departure, I conceptualize African subjectivity here through processes of migration, encounter, and circulation (of peoples, ideas, commodities). These are similar processes that mark Black subjectivity in the Western Hemisphere, hence "African-Americans are tied more closely to their African heritage not because they can be traced through direct lineage to a particular people or area of Africa, but because they, like Africans, are products of different peoples coming together through historical processes" (Gregg 2001, 266–267). Tsitsi Jaji (2014), Jemima Pierre (2008, 2012), and Hershini Bhana Young (2006) all argue for more nuanced discussions of contemporary African experiences in relation to diaspora studies. In general, they lament the lack of coevalness (Fabian 2002) with which African experiences are analyzed within African Diaspora studies. By insisting on the continued and multivalent processes of subjectivity, the concept of Afrodiasporic Space counters the tendency to frame Africa in the past where its importance to diaspora theorizing is too often rooted. In Afrodiasporic Space, Africa is a constitutive and continuous site of diaspora. This is not in the service of inclusion or affirmation, although these are not unimportant goals, but for the purposes of rethinking the kinds of interventions African Diaspora theory might make, and how practices of solidarity and freedom might be reimagined.

Second, Afrodiasporic Space also reimagines diasporas as theoretical concepts. Avtar Brah (1996) proposes the idea of "Diaspora Space" to help define the concept of diaspora. Diaspora Space is a conceptual category that is "inhabited not only by those who have migrated and their descendants, but equally by those who are constructed and represented as indigenous" (209). This concept is useful when we imagine the African Diaspora within the configuration that Brah outlines as Diaspora Space, because it suggests that the African continent is as much constructed by the discourses of the African Diaspora as the diasporic spaces are. In addition, Africa is a diasporic space itself. As such, Africa and its diasporas inherit not the same physical space but the same conceptual and ideological space, which then forms the basis of knowledge production and intellectual inquiry. For Brah, the issue with the study of diaspora is not one of historicizing or addressing the specificity of particular diasporic experiences. She acknowledges that this is important but imagining Afrodiasporic Space as an analytic concept requires that what is attended to is "the historically variable forms of relationality within and between diasporas" (Brah 1996, 183). This allows us to account for the differently lived experiences in Afrodiasporic Space as well as the contestations within various diasporic sites that may arise out of race, gender, sexuality, class, and nationality.

Looking at Afrodiasporic Space through intersectional forms of materiality informs the meaning of kwaito bodies. To examine which bodies are central to how the nation is performed I develop the concept of "kwaito bodies." Kwaito bodies denotes the generation of Black people who came into adolescence and adulthood in the years following the demise of apartheid. This generation created the set of musical and cultural practices known as kwaito. Referred to colloquially as the Y generation, they were the first group of Black people to take advantage of the changes created by the new political dispensation. Thus, this group was fundamentally important in defining the possibilities of post-apartheid South Africa for future generations of Black bodies.

Afrodiasporic Space and kwaito bodies are analytical tools that allow me to critically engage contemporary practices of subjectivity in South Africa. Examining kwaito bodies as Afrodiasporic reveals how kwaito adherents imagine both an alternative formation of the national body as well as an alternative conceptualization of freedom for post-apartheid South Africa. Furthermore, by engaging contemporary cultural practices within the African continent, the very definition of diaspora shifts from historical, linear flows common to much of diaspora theorizing and moves into an understanding of diaspora as processual, circulatory, and polyphonic. Ultimately, considering the diasporic nature of contemporary African cultural practices such as kwaito reimagines both the study of contemporary (South) Africa as well as diaspora studies itself.

My own co-performative body, Beyoncé's "Run the World (Girls)" video, and Boom Shaka's performance of "Nkosi Sikelel' iAfrika" are all placed into conversation in order to illuminate contemporary circulations of global Black cultures in which kwaito participates and to show how kwaito adherents operate in and through Afrodiasporic Space. In each case, we see how kwaito operates as a site for the creative combination and reworking of Afrodiasporic cultures both within and outside of the continent. We also glimpse how diaspora is a product of particular kinds of material exchange that involve the movement of bodies as well as the movement of cultural products both across the Atlantic and *within Africa itself*. "Kwaito performances and music videos give momentum to an imaginary community or nation ([Benedict] Anderson) by combining styles and symbols of local Black history with those of the contemporary Black Atlantic" (Pietilä 2013, 2). Kwaito performance suggests alternative formations of the nation, formations that reveal South Africa to be the product of numerous transnational encounters. As a result, the nation is not an assumed insular entity; instead, it is already scripted as a global, transnational project. Furthermore, this performance of the nation imagines young Black people, many of whom live in the townships, as the active subjects of post-apartheid freedom.

Misrecognition and/as African Diaspora

Identifying South Africa as Afrodiasporic Space does not imply that relationships among diasporic spaces are always harmonious. On the contrary, Afrodiaspora often coalesces around different forms of misrecognition that enable possibilities for connection and solidarity. Describing an encounter from Langston Hughes's travels to Liberia, Kenneth Warren (1993, 400) reveals "this comedy of misrecognition, in which Hughes, who appeals for misrecognition as an African is misrecognized as white and George who appeals for misrecognition as a Kentuckian is misrecognized as African points to the condition of the diasporan subject. Recognition can never preclude misrecognition because one can always be identified as other than what one claims to be." Inhabiting the spaces of ambiguity, incompleteness, and (un)knowing is necessary for what Brent Hayes Edwards (2003) calls the "practice of diaspora." (Mis)recognition can be productive; while it is often a site for the reproduction of social inequality, moments of misrecognition also are opportunities for critique. Misrecognition often reveals both subtle and explicit matrices of power. Writing about her own experiences of being misrecognized as straight due to her femme gender performance, Kaila Story (2017) argues that this misrecognition gives her the opportunity to critique (Black) heteropatriarchy. (Mis)recognition as diaspora inhabits the space of "friction" that may exist between Black people in Afrodiasporic spaces while also insisting that such friction can be productive spaces of affinity.

The Afrodiasporic Space analyzed is one that I, a queer African American researcher, exist in as well as create. As a dark-skinned African American man who could easily pass for West African, a sometimes obvious Black "other," I was often subjected to additional police scrutiny and asked to prove my right to be in the country legally. Checkpoints and police surveillance seemed omnipresent in communities in which I traveled, particularly before major holidays and at month's end when many people in South Africa receive their paychecks. During one Christmas holiday, I was pulled over by the police at a checkpoint. I showed my California driver's license, which, according to South African law, allows international visitors to drive legally in the country. Despite this, I was asked to produce an international driver's license, which I did not actually need. When I pressed my point with the police officers, I was arrested and my car confiscated.

Initially, I thought that I was being arrested for driving without an appropriate license, but I was shocked to discover upon my arrival at the police station that I was being booked as an illegal immigrant. This experience was one I can imagine few other non-Black American researchers were subject to in South Africa. Once arrested and taken to the police station, I pleaded with the officers to allow me to go to my home to produce proof of my legal residence; they refused. This was despite my home being a mere five-minute drive from the checkpoint. Instead, I was jailed along with the other "illegal immigrants" who had been captured. Significantly, the arresting officers as well as those in command were Black South Africans, and all of the "illegal immigrants" with whom I was booked were African nationals from Zimbabwe.

On this day I had failed to perform my American identity sufficiently to avoid arrest. And lacking papers, I was simply part of an undifferentiated mass of *makwerekwere*, a derogatory term used to describe African immigrants to South Africa. I was also unable to sufficiently perform South Africanness. A friend of mine described linguistic performance as central to convincing the police that one is a Black South African. Not only must one speak an indigenous South African language, but that discourse should preferably not be in Zulu, since Zulu serves as a lingua franca that is often picked up by non–South Africans. Xhosa is the preferred (South) African language because it is the only major African language not spoken widely in one of South Africa's neighboring countries and thus proves a certain autochthony. Apparently even shouting a few words in Afrikaans, a language few if any Black foreigners bother to learn, can substantiate its Black speaker as South African. Lighter skin is also preferable; numerous dark-skinned South Africans have been subjected to arrests and threats of deportation (Vigneswaran 2007). In this way, South Africa continues to function as a "white" country since I could find no evidence of whites being stopped at these checkpoints. My fellow white American researchers certainly were never required to prove they were in the country legally.

I could take some solace in the fact that, for whatever reason, illegal immigrants were processed separately from those arrested for other crimes. I was, however, appalled that one of my cellmates appeared to be a child and should not have been kept in the cell with us. During this time, my Zimbabwean compatriots soothed themselves with prayer and song while I plotted how I was to avoid being shipped to the dreaded Lindela facility where people have literally disappeared and where numerous human and civil rights violations have been documented (Guma 2012; Sutton et al. 2011). I would like to say that the US embassy was helpful in this matter, but they were not. They sent a fax, verifying my legal status in the country—a fax that was ignored for hours by those at the police station. Ultimately it was a Black queer South African friend whose partner was in prison who was able to free me. He explained later that he has a lot of experience dealing with the police and was able to convince them to allow him to see me. He made it clear that I was arrested more for my attitude—for daring to contradict police authority—than for any violation of the law. The presiding captain refused to allow me to return home to produce my passport since I had such a bad attitude. Once I gave my friend my keys, he went to my home and produced my passport, thus freeing me from a nearly ten-hour confinement.

Upon my release, I accepted the responsibility of phoning the friends and family of my Zimbabwean cellmates, many of whom were not allowed the courtesy of a phone call (so perhaps my Americanness was helpful in some other ways on this day). When I phoned the girlfriend of one of my cellmates, she seemed unsurprised by his arrest and simply asked me how much bribe money she would have to come up with in order to release her boyfriend. Her voice sounded resigned, as if she had done this on a number of occasions. I actually had absolutely no idea how much money the police would want for his release, but I told her 300 rand (at that time about US$50), hoping that this would give her some small reassurance. I do not know what happened to any of the men I was in the cell with that day, whether any of them were able to bribe their way out, or what happened to them once they were transported to Lindela. I made a point from that moment forward to never leave home without my passport on me. I should reiterate that the burden of having to constantly travel with a passport was never one borne or even considered by my white American counterparts.

By the time I completed fieldwork in South Africa, I counted at least ten stops by the police. The experience became so commonplace that I did not question it. I simply knew that I would be pulled over if I were ever at a checkpoint and that I could expect that I might be pulled over while driving in the communities in which I lived and did research. Although I always had my passport with me, I learned that carrying my passport did not always provide me with protection. Even when I had my passport on me, I was typically subjected to additional scrutiny. At one police stop after producing my passport with its valid visa, I was asked incredulously about my place of birth. It was as if being a naturalized American citizen (as he suspected) would somehow have made me less of an American, or perhaps less entitled to the deference that Americans are typically given in South African society. At these checkpoints with predominantly Black South African police officers, I would be told that I did not look American. To which I often replied, "What does an American look like?" Apparently, not like me.

Compounding these moments of state (mis)recognition are moments of everyday (mis)recognition in my relations with Black South Africans. I have also had to explain to some Black South Africans why I do not "sound Black," why I cannot rap, and why I do not play basketball. After saying that I am from California, I have to explain that I did not grow up in Compton or South Central LA or the "projects"—ideas that seem to dominate mass-mediated imaginaries of African American experience, often due to the popularity of a variety of US reality television shows featuring African Americans. My illegibility in the moments that I describe is a central component of Afrodiasporic Space. In those moments where I was unreadable as (African) American, subjected to a different kind of state surveillance as a "Black other," and (mis)recognized even when I was read as African American, I was provided an opportunity for revelation and newfound affinities with other Black subjectivities, both queer and nonqueer.

In being illegible as a particular type of Black subject, the complexity of transnational Blackness is misrecognized—for in this instance I can only be a Black illegal, nothing more. This misrecognition is also rooted in the misrecognition of Black genders and sexualities, for my performance of gender and sexuality, my lack of African American (hyper)masculinity, also marked me as illegible. Rather than dwell in this illegibility, however, these patterns of misrecognition provided ample space for critique and solidarity. While ephemeral, my experience as "illegal" in South Africa (and the constant threat of having to prove my belonging) allowed me to create forms of solidarity that critiqued immigration laws rooted in white supremacy. I have also been able to critically reflect on the classed nature of these experiences. As my own income and status rose from that of broke graduate student to more financially secure middle-class professor, I found myself able to increasingly circumvent the quotidian experiences of police harassment. My experience of nonnormative gender and sexuality also provided a space to critique both the limited circulations of Black life in global media industries and the assumptions of gender and sexuality performance that accompany such circulations that are rooted in Black heteropatriarchy in South Africa. It is these kinds of possible interactions even within (mis)recognition that are made possible through Afrodiasporic Space and that are mobilized in a series of performances that I highlight.

Traveling Bodies: Kwaito on the Move

In 2011, Beyoncé, the international R&B superstar turned global brand, released her fourth studio album 4, featuring the song "Run the World (Girls)." The music video was widely celebrated for its choreographic blending of elements of hip-hop, dancehall, and African movement. Despite her popularity, I had paid little attention to the hype surrounding her latest album and had consciously chosen to avoid the numerous media appearances, live performances, and links to her songs and videos. Considering her iconic status in global Black culture, by the time I arrived in South Africa for a short research trip in June 2011, I had not engaged with any of Beyoncé's most recent material. Fortunately for me, Beyoncé has just as rabid a following among Black gay men in South Africa as she does in the United States. During one of my conversations with a young Black South African gay man, I learned that Beyoncé's video for "Run the World (Girls)" might prove to be a more interesting performative text than I could have imagined as he pointed out to me that the video for the song features a number of popular kwaito-styled dances. "Run the World (Girls)" in fact demonstrates the contemporary workings of Afrodiasporic exchange that mark the global circulation of kwaito. The central (mis)recognition operating here, much like the example cited by Jaji earlier in the chapter, is the equivalence of forms of diasporic intimacy with cultural imperialism.

Beyoncé has received accolades from a number of music critics for bringing complex choreography back into music videos. Her up-tempo songs ranging from "Crazy in Love" to "Single Ladies" and the more recent "Run the World (Girls)" and "Formation" have spawned numerous YouTube videos of fans attempting to replicate the energetic and sensual dance moves. Flash mobs in cities as diverse as London and Ljubljana have routinely performed the choreography from her music videos. Likewise, she has been critiqued for stealing choreography from a variety of cultural sources, without always giving proper attribution. However, the process through which Beyoncé showcases kwaito-oriented dances to a global audience reveals contemporary diasporic circulations whose meanings are far more complex than mere cultural imperialism enabled by new media. Instead, her video for "Run the World (Girls)," particularly its choreography, demonstrates the usefulness of a diasporic critique central to understanding kwaito performance.

According to Beyoncé, the song's beat drove her interest in making a movement-centered video. "Run the World (Girls)" relies heavily on samples of "Pon the Floor," a dancehall- and funk carioca–inflected song released in 2009. There are also elements of go-go and African percussion. In a short film documenting the year-long process of crafting the album, Beyoncé speaks about the percussive elements of the song, stating that the beat "takes you back to Africa" (Burke 2011). Later in the documentary, Beyoncé speaks of seeing a YouTube video featuring the *pantsula* dance troupe Tofo Tofo performing at a Mozambican wedding. It is unclear whether Beyoncé herself discovered the video or was introduced to it by her choreographer, Frank Gaston. Beyoncé's choreographers are (in)famous for scouring YouTube videos for source material for her dance moves, a practice that has not always been welcome in performance communities, particularly when the source material remains unacknowledged. For "Single Ladies," Beyoncé's choreographers imaginatively reconfigured Bob Fosse's "Mexican Breakfast" routine. What remained unacknowledged was that a YouTube video preceding Beyoncé's had existed for some time that matched Fosse's choreography to the Atlanta hip-hop artist Unk's song "Walk It Out," and it undoubtedly had provided some inspiration for the reuse of this particular choreographic routine in US-based urban music. This is to say nothing of the incorporation of vogueing and j-setting elements riffed from US-based Black queer performance repertoires that also go unacknowledged and the more recent concerns of independent artists from New Orleans who allege that Beyoncé's album *Lemonade* borrows too heavily from their own filmic and musical projects.

Beyoncé recalls viewing the YouTube performance of Tofo Tofo a year prior to the release of her *4* album and deciding that she would incorporate this choreography into her next up-tempo song. Her choreographer Frank Gaston explains, "We had seen something on YouTube; we had seen these three guys from Africa, this Mozambique African dance troupe. . . . [W]e were like, 'Wow this is an amazing movement.' . . . And that movement has always been at the back of our head for the last year" (Vena 2011). The filmic documentary makes much of the fact that Beyoncé and her team of expert dancers and choreographers struggled with the kwaito-styled dancing of Tofo Tofo. Beyoncé suggests that Western-trained dancers and choreographers were sanitizing the movement, making it "too perfect" and losing the joy encapsulated in the YouTube video. She opines, "I haven't seen one person do it right. . . . [N]one of the top dancers . . . the top choreographers could replicate this dance" (Burke 2011). As a result, Beyoncé decided that Tofo Tofo would have to be flown to Los Angeles in order to teach the choreography to the singer, her choreographers, and the professional dancers hired for the video shoot. One does wonder, however, if she and her dancers had been able to replicate the movement to her satisfaction, whether anyone would have bothered to acknowledge Tofo Tofo's contribution at all or if the video would have continued a long-standing trend of major stars mining cultural practices from marginalized spaces with little or no citation of the source material. Thankfully in this case, Beyoncé and her choreographers give due credit to Tofo Tofo:

> While Beyoncé is the pop star, Gaston is quick to give credit to all of the dancers in the clip, including some that the team enlisted from Africa. "We say, 'Share the light, you sparkle brighter,' and we share our light with a lot of new, creative people. I feel like we really nailed it and, again, my hat's off to the Tofo Tofo guys [from Africa], 'cause none of us could imitate that," he said. "We had to bring them around to learn that [move], which is really, really interesting. They had such a great vocabulary of movement. Those two dancers not only helped shape the moves in the video, but also moved Beyoncé. That was probably one of the most beautiful experiences for Beyoncé. They were so humbled," he said. "It was hard finding them. They were really in a remote area; we had to get the embassy people involved. That was a process that took about two months or more. Beyoncé really loved them and I'm pretty sure we'll see them again. It was magical." (Vena 2011)

Unfortunately, the dancers of Tofo Tofo are portrayed as authentic, unspoiled Africans who exist in some primordial, rural past that makes them difficult to locate and thus necessitated over two months of diligent searching and the intervention of US embassy personnel in Mozambique in order to find them. This discursive positioning of African peoples and their cultures in a primordial past by contemporary new-world African descendants is a phenomenon that is not limited to Beyoncé's filmic text. Instead, the positioning of Africa as origin-past to African Diaspora people, cultures, and practices permeates much theorizing in African Diaspora studies. Contemporary African-based cultural practices such as kwaito require a reimagining of this limiting analytic.

First, we have to consider the way in which Beyoncé and her choreographers came to know about Tofo Tofo in the first place. YouTube served as their introduction. This fact alone disturbs the characterization of Tofo Tofo as divorced from postmodern cultural flows. While it is the case that Tofo Tofo did not produce the video or post it on YouTube, that their performance appears through this medium suggests that its members are more directly connected to contemporary global cultural practices than is imagined by the documentary. Even if Tofo Tofo are firmly ensconced in the rural hinterlands of Mozambique, they are not disconnected from new media practices.

In an interview with the group conducted by the South African newspaper *City Press*, the group members chart their journey from street dancers in Maputo, Mozambique, to being featured in Beyoncé's video. They stated that they were approached via a promoter in London who had contacts with a Maputo-based dancer. This dancer then contacted the members of Tofo Tofo, informing them that Beyoncé wanted to work with them on her latest video project. Mario Buce, one of the members of the group, admitted, "We couldn't believe it" (Seabi 2011). According to the article, the group members have a popular local following that predates their working with Beyoncé: "the dancers are popular in Mozambique and are always in demand" (Seabi 2011). At one point in the documentary, the members of the group are portrayed as if they did not know who Beyoncé was. After Beyoncé greets the group members, one of them asks, "What's your name?" and she replies, "I'm Beyoncé" (Burke 2011). It is not clear if the members of the group initiated this exchange or if they were asked to do so by Beyoncé's production team. However, it is unlikely that individuals living in the global metropolis that is Maputo (with a population of 1.76 million as of the 2007 census), no matter how impoverished, would not know of the international pop icon. It seems as if this was another move on the part of the filmic text to portray them in a primordial, "authentically African" past and thus removed from global cultural flows. The *City Press* article does mention that the group members are from impoverished backgrounds and have little access to the internet, making it difficult for them to watch their own performances online, but, significantly, the article makes no mention of the group members being rural-based.

Even if we accept the filmic text at face value, the types of dances employed by Tofo Tofo would complicate any narrative that situates them outside of global Black cultural flows. Tofo Tofo primarily incorporates pantsula-style dancing into their routines. Pantsula refers to a range of cultural practices that emerged in South African townships and was originally associated with tsotsi (gangster) culture of the 1950s. One pantsula dancer describes the culture as "more than a dance, it is a way of life, an art that he breathes" (Moore 2009): "Pantsula evolved on the streets of South Africa just as hip-hop developed in American ghettos. In South African townships, pantsula was used to refer to someone who dressed elegantly yet informally with a degree of panache. The dance form really grew during the 1980s when young men in the townships discovered it as a means of self-expression and took to the dance with enthusiasm."

It is not coincidental that many proponents of pantsula dance connect its social function and rise to hip-hop, which performed a similar role in US-based Black communities in the 1980s. These were dances literally performed "in the streets" with major battles and competitions between crews. When kwaito emerged in the 1990s, it became the preferred soundtrack for pantsula dancing and competition with several prominent kwaito groups (among them Mashamplani and Trompies) explicitly incorporating pantsula style and dance into their images. The spread of pantsula style and dancing to other Southern African locales, including Mozambique, likely was a product of the same kinds of migrant labor circulation that spread various cultural and religious practices that buttressed other forms of popular culture, such as choral-style singing throughout Southern and Eastern Africa.

Mozambicans would also be familiar with pantsula-style dancing from their exposure to forms of South African media that circulated in postwar Mozambique. *Makwaela*, a form of dance that was promoted by the post-independence Mozambican government as an example of revolutionary culture, is heavily influenced by practices reconfigured from migrant laborers who developed hybrid cultural forms in concert with Africans from a variety of backgrounds who were employed in the South African mines. The connection with migrant labor also reveals the gendered aspects of this type of choreography. Pantsula-style dancing with its connection to masculinity and migrant labor is a heavily gendered form of performance that tends to be limited to men. Given its association with the township, this form of dance has a cachet that links it to authentic Black masculine performance, but as a style of performance it is hardly hegemonic. Pantsula dance receives little funding or support from the arts community, which tends to prioritize high-art Western-based dance forms first, followed by "traditional" forms of dance.

The sight of Beyoncé, who often represents a type of global Black feminine ideal, wearing haute couture and performing these dances is somewhat dissonant with the practice given that South African women who dance pantsula struggle for recognition and opportunity (see "Two Women" 2017). Hence, we might consider her adoption of these styles of dance as a form of diasporic (mis)recognition. Yet it may also reveal the ways in which her body as an (African) American and a global superstar allows her to transcend the gendered divisions that mark this style of dance. Based on my discussions with friends and my analysis of media commentary regarding Beyoncé's engagement with pantsula, the gendered dimension of her performance went unremarked. By performing these dances, she opens them up to gendered reworking. Additionally, the Tofo Tofo dancers were also restyled for the video apparently wearing similar designer-styled clothing, striking another note of dissonance. It is almost as if the very sanitation that Beyoncé did not desire for the movement had been transferred to the sartorial, as pantsula dancing generally involves colorfully restyled workwear, not haute couture. In the one homage to the

FIGURE 1.1. Beyoncé dancing with Tofo Tofo in "Run the World (Girls)." Provided courtesy of Parkwood Entertainment, Chokolate LLC, and Columbia Records.

FIGURE 1.2. Beyoncé poses with Tofo Tofo. Provided courtesy of Parkwood Entertainment, Chokolate LLC, and Columbia Records.

FIGURE 1.3. A depiction of pantsula dance movement and style. Source: Tyrone Bradley.

tradition of pantsula-style dance and sartorial practices, the dancers did retain their Converse sneakers, which are quickly flashed as the camera focuses on their complicated footwork (see figures 1.1–1.3).

Boom Shaka Performs "Nkosi Sikelel' iAfrika"

In May 1998, Lebo Mathosa, Theo Nhlengethwa, Thembi Seete, and Junior Sokhela—the members of the kwaito group Boom Shaka—created much controversy with their performance of the South African liberation song / national anthem "Nkosi Sikelel' iAfrika" during the South African Music Awards (SAMAS). I focus on their performance because it marks an early and important example of how Afrodiasporic Space operates in post-apartheid South Africa, and the possibilities of Afrodiasporic analysis for rethinking South African cultural criticism. It is also a critical example of the (mis)recognitions that permeate Afrodiasporic Space, given that critics of Boom Shaka's performance misrecognized it as un–South African for both its "outside" influences and its nonnormative performances of gender and sexuality.

FIGURE 1.4. Boom Shaka group members (*left to right*): Thembi Seete, Junior Sokhela, Lebo Mathosa, and Theo Nhlengethwa. Source: www.all4wineb.co.za.

I viewed a brief clip of the performance in 2001 as part of a larger documentary about the South African national anthem. I was struck, although not surprised, by the centrality of women's bodies to this rendition of "Nkosi," even though almost all of the media commentary on the event had noted this foregrounded presence. Importantly, women's bodies and voices have been key to the performance tradition of "Nkosi" (Redmond 2013, 634–635). Charl Blignaut (1998) describes Boom Shaka's performance in a newspaper article: "Stylishly clad in the deepest of blue velvet suits over lacy bras and flimsy white blouses held in place by at least one button—Boom Shaka's Thembi and Lebo (the two women members of the group) had walked slowly to the front of the large Civic Theatre Stage and then stopped, each raising a clenched fist in the air. A pounding beat kicked in, sending a wave of motion down the girl's bodies." Equally focused on the women of the group, another journalist (Pokwana 1998) also describes the performance: "Free from the usual groups of males with a banner exhorting the two sassy and sexy girls in the group to take their tops off, the group executed the anthem with supple energy coupled with flexible choreography, leaving those with stiffer waists aghast."

Vukile Pokwana's observations suggest that the power of the performance was due not only to its unbridled sensuality but also to the ways in which the sensuality appeared divorced from the demands of masculine pleasure. Much

Afrodiasporic Space 43

FIGURE 1.5. Lebo Mathosa (*left*) and Thembi Seete. Source: Media24.

of the public discomfort with Boom Shaka's performance is a reaction to the group's decision to center erotic embodiment—especially by Black women—in performance. Thus, this controversy must be viewed in a larger context in which Boom Shaka's performances are already understood as contravening social mores, particularly the highly stylized, sexually charged performances of its female participants, Lebo Mathosa and Thembi Seete (see figures 1.4–1.5).

In 1998, the SAMAs were a much smaller, far less lavish affair than they are now. Today the awards show is broadcast live from the 6,000-seat Sun City Superbowl and the ceremony itself is just one event in a weekend full of parties, performances, and networking for the South African music industry. In 1998, the SAMAs were held in the Civic Theatre of Johannesburg, a venue that seats only 1,100 people at full capacity. More importantly, the awards show at that time was not broadcast live. As a result, very few people saw Boom Shaka's performance of "Nkosi" in the moment. The performance and subsequent analysis of its meaning for post-apartheid South Africa were, in the words of Diana Taylor (2003), "archival," in that an array of mediated material existed to frame the public discussion. Writing in the *Mail and Guardian* several weeks after the fact, Blignaut observed that "the somewhat dismal taping of the awards . . . failed, almost spectacularly, to convey any of the energy of what had been a breakthrough year at the SAMAS" (1998). Thus, the media were left to interpret the apparent controversy that ensued after the performance with little direct public knowledge of the event.

I did not attend the performance that was the source of the controversy, and I did not view the taping of the awards ceremony in its immediate rebroadcast. But even two decades later, this "Nkosi" affair was still mentioned frequently in media reports about the members of Boom Shaka. Most critiques of Boom Shaka's performance suggest that their version was disrespectful to the spirit of "Nkosi" in its twin role as a liberation song and new national anthem. The articles used words such as "immoral," "degrading," "lacking in respect," and "unacceptable" (*Mail and Guardian* 1998). Implicit in this aversion is the fear that something about the anthem has been tarnished by the performance. Originally South Africa's Black national anthem, "Nkosi" was blended with the apartheid anthem "Die Stem" (given new English lyrics for the final stanza) to become the post-apartheid South African national anthem. Shana Redmond's important study (2013) notes that while the song has traditionally been performed in a "slow, meditative performance tradition" (631), Boom Shaka's performance sped the song up to a kwaito beat, moving the song from solemnity to celebration while also using experimentation to represent the unfinished business of liberation.

"Nkosi" might be unique among Black anthems, given that this twin role and the tension between its original outsider/minoritarian status and its official status as national anthem mirror the repositioning of Blackness in the post-apartheid state. Redmond notes that "Nkosi," like many black anthemic songs throughout history, has played a strategic role in creating unity across difference and fostering rebellion. Redmond distinguishes between Black anthems and national anthems: "Unlike standard national anthems Black anthems were not ubiquitous but instead were performed selectively, and even when their usage was not formalized there was always some clarity of *ends* engendered by the performance" (2013, 26, emphasis original). But what happens, as is the case with "Nkosi," when the Black anthem becomes the national anthem, when it moves from a space of rebellion to state-sanctioned approval? When its mode shifts from anticipatory to triumphant? Are the flexibility and dynamism of performing the song jettisoned in favor of static statist interests? Redmond argues that the shift to incorporation as part of its becoming the South African national anthem reveals the violence of the state and the ways in which racialized colonial violence are "re(entrenched) through music even in the pursuit and wake of racial justice victories" (2013, 55). Boom Shaka uses the performance of "Nkosi" to disrupt the ultimately violent narrative of the inclusive, collegial post-apartheid state, in order to perform an alternative formation. Much like historical performances of "Nkosi," Boom Shaka's interpretation effectively performed their citizenship. Through their performance, Boom Shaka insisted that the state be enacted more inclusively, pushing against its heteropatriarchal formation. In speeding up the overall tempo of the song and performing it in an aurally unrecognizable register, Boom Shaka appears to move past the moment of triumph and offer an almost chaotic rendering of the anthem. The faster tempo performs the labor of simultaneously marking the moment of achievement of moving the song from solemnity to celebration while also, in its chaotic unfamiliar rendering, revealing the unfinished business of liberation. Hence, the aural register of Boom Shaka's version of "Nkosi" refuses to dwell in the moment of triumph through liberation and instead begins to ask questions about the practices of freedom. Boom Shaka's performance asks difficult questions about exactly whose interests the new post-apartheid state will serve.

Sandile Ngidi, a South African journalist, writer, and literary agent, argues in his discussion of the performance that Boom Shaka's rendition—because of its "erotic" nature—caught the nation "with [its] pants down" (Madondo 1998). If we take Ngidi's assertion very seriously, we can expand his definition of the erotic beyond an immediate connection with the sexual and think about the other pleasures that the erotic evinces. Audre Lorde's (1984) conceptualization of the erotic as a deeply spiritual, feminine power often denied by Westocentric masculine discourses is helpful here, particularly her commentary on the erotics of sound. Connecting the experience of joy specifically with music and rhythm, the erotic surfaces "in the way [the] body stretches to music and opens into response harkening to its deepest rhythms so every level upon which I sense also opens to the erotically satisfying experience" (Lorde 1984, 36). The erotic not only can be used to think about this singular performance and the controversy surrounding it but also can be used as a rhythm by which to examine how kwaito performances more generally assist in hearing the future of the post-apartheid nation. What does it mean for young Black bodies—infused with energetic joy and unbridled sensual pleasure—to perform the nation? If Bhekizizwe Peterson is correct and the problem for critics of Boom Shaka's performance is that "national anthems are no dancing matter" (2003, 206), in whose interests is the joyful celebration of dance disconnected from the celebration of both national unity and the liberation struggle? For many in the public domain, the central controversy of the performance has to do with appropriateness as it relates to young bodies, to young Black bodies, and to young Black women's bodies. While the specific focus is on Boom Shaka's performance of a singular song, "Nkosi Sikelel' iAfrika," the larger public debate is over the meaning of the bodies performing the song. Thus, it is crucial to reorient the debate in light of one of the central questions that animate this study: which bodies are being critiqued and why? How does the misrecognition of transnational Blackness (in Afrodiasporic Space) and alternative performances of gender and sexuality contribute to the misreading of this moment?

Boom Shaka's performance of "Nkosi" is simply one example of how kwaito bodies cannot be contained by the national script of Black heteropatriarchal redemption. Significantly, kwaito bodies become illegible if national belonging is defined by purity and an implicit autochthony in both its pre- and post-apartheid guises. Kwaito performers engage, extract, and create their musical practices from a complex array of Afrodiasporic entanglements. Several aspects of Boom Shaka's performance of "Nkosi" could be said to draw on forms of globally circulating Afrodiasporic practices. Four aspects of the performance will be examined: the dancing exhibited by the group members (particularly Lebo Mathosa and Thembi Seete), the ragga chanting of Junior Sokhela, the backing house track for the song, and the composition "Nkosi" itself. While "Nkosi" has been adopted as the song of the South African liberation struggle, akin in many ways to "We Shall Overcome" in the United States, the "original" "Nkosi" was a product of Afrodiasporic encounters.

First, Boom Shaka's performance was described as erotic, funky, and sexually charged due to the very sensual nature of dance moves that emphasized lithe, dexterous, winding movements and pelvic articulations. To be clear, such dance moves are not absent from South African performance practices generally, but the choreography enacted by Boom Shaka owed a considerable debt to dances borrowed and rearticulated from both Jamaican dancehall and Congolese dances, as Jamaican dancehall scholar Sonjah Stanley Niaah (2010) notes. Sensual, lithe movements borrowed and rearticulated from 1990s dancehall queens such as Patra found their way into the movements and choreography of several women-centered kwaito groups, including Boom Shaka. Indeed, kinesthetically as much as sartorially, Boom Shaka's women owed much to dancehall queens with short shorts that rode high on the thighs (known in dancehall circles as *pum-pum* shorts). Furthermore, the ladies of Boom Shaka were also known for wearing the long, thick braids common to many dancehall queens of the early to mid-1990s. In fact, this style of hair braiding became so closely associated with Boom Shaka that it was (and continues to be) popularly referred to as Boom Shaka braids. While such sartorial practices were quite popular although not exclusive to Jamaican dancehall, they were not common aspects of women's sartorial regimes in South Africa and must have been a shocking sight to traditionalists and those of bourgeois sensibility alike. "Wining," a form of sensual, female-centered dance central to the dancehall aesthetic, was appropriated and reworked by the women of Boom Shaka to great effect.

Additionally, the pelvic and hip-centric movements of Congolese dance, particularly the *ndombolo* and *kwassa-kwassa* styles, were incorporated into the movement of both the women and men of Boom Shaka. As South Africa opened its post-apartheid borders, large numbers of Congolese migrants began to settle in the Johannesburg area, congregating in the very neighborhoods that were first opening up to Black South Africans where kwaito was incubated. These new migrants were present and influential in many of the nightlife spaces that fostered kwaito. In addition, tapes featuring performances and music videos by Congolese artists began to circulate in post-apartheid South Africa. Subsequently, many of these dance crazes were reappropriated by South African kwaito musicians (see Peterson 2003).

Second, the ragga-muffin chanting of Boom Shaka's Junior Sokhela also represented an engagement with Afrodiasporic culture that pointed toward the Caribbean. Much like Congolese music, Jamaican dancehall music was also spreading in popularity throughout post-apartheid South Africa. The "toasting" style symbolic of Jamaican dancehall, which was popularized by artists such as Shabba Ranks, was adopted by many kwaito groups that featured one male member who chanted in this style. Not only did these artists tend to reconfigure Jamaican toasting lyricism, they also often used the gravelly, gruff-voiced tones popularized by artists such as Ranks and Buju Banton yet also present in the vocal tones of local artists such as Mahlatini. Dancehall was popular in the nightclubs in mid-1990s Johannesburg, particularly in Yeoville clubs such as Tandoor, and dancehall enjoyed sonic parity with R&B, hip-hop, kwaito, and house music. However, more recently, dancehall has been mostly driven (temporarily?) underground and marginalized. Even the popularity of the term "ragamuffin" to describe Jamaican dancehall points to diasporic influences, as "ragga(muffin)" was a term that was popular in Black British communities initially to describe a particular subgenre of dancehall music that morphed into an umbrella term for dancehall in Britain. Much of the dancehall in South Africa circulated because of South Africa's relationships with Britain and Black

British culture and aesthetics. In "Nkosi," Junior appropriates the ragga toast to praise heroes of the struggle, including Nelson Mandela and Steve Biko.

The third aspect of the performance that points to its Afrodiasporic imprint is its sonic texture, which relies heavily on a backing house track to create its rhythm. As Gavin Steingo explains, "Usually the compositional process in kwaito involves slowing down House tracks (or simply creating 'slowed down' House tracks) and then adding vocals and samples from a variety of sources. Most often, the House tracks are looped and repetitive phrases are chanted: usually in Zulu, Sesotho, English, or kwaito-speak. . . . While there is a fairly wide variety of kwaito today, Boom Shaka's version of Nkosi . . . fits into the predominant productive processes . . . outlined" (2008b, 109).

I am interested not so much in demarcating the lines between kwaito and house music, nor suggesting a particular historicization that attempts to mark the sonic differences between the two genres within South Africa or the specific shifts that occurred when house music became known as kwaito. Rather, I am interested in post-Fordist global circulations of Black youth musics and cultures that are not confined to hip-hop. I also want to point out some of the sociopolitical similarities between the rise of house music in the Midwestern United States and its ultimate embrace by Black South African youth. As Steingo notes, "In many ways, kwaito is the expression of millions of Black youths who are trying to make sense of a violently globalized liberal-democratic landscape" (2008b, 106).

Writing about Chicago house music, Anthony Thomas makes a compelling argument for the ways in which the genealogy of house music has been disremembered. He states:

> America's critical establishment has yet to acknowledge the contributions made by gay Afro-Americans. Yet Black (and often white) society continues to adopt cultural and social patterns from the gay Black subculture. . . . What's also continued to emerge from the underground is the dance music of gay Black America. More energetic and polyrhythmic than the sensibility of straight African Americans, and simply more African than the sensibility of white gays, the musical sensibility of today's "house" music—like that of disco and club music before it—has spread beyond the gay Black subculture to influence broader musical tastes. (1995, 437)

My aim is not to summarily dismiss scholarly interpretations of house music that seek to qualify or destabilize some of Thomas's claims concerning the genealogy of house (for a good analysis, see Fikentscher 2000). However, I do share Thomas's concern for remembering the role of African Americans in the creation of global house music and in paying specific attention to the politics of gender and sexuality in the formation of these musical genres. The circulation and adoption of house music in South Africa is part of what DJ/scholar Lynnée Denise calls an "Afro Digital Migration" (2012). This migration though does not posit the United States as the singular point of origin and South Africa as the singular receiving space. Instead, the kinds of global exchanges that are characteristic within house music and exhibited through kwaito reposition South Africa and the United States as two possible nodes within a larger spatial framework that I refer to as Afrodiasporic Space.

As discussed by South African journalist Greg Bowes (2015), kwaito was firmly entrenched in the trends of international globally circulating house music. For Bowes, kwaito was essentially "a combination of house's rhythmic engine (in second gear) and colloquial commentary.... [I]t was international house music cut with the colloquial vocal sentiments of an optimistic South Africa that just wanted to feel liberated." Kwaito was spawned by the local popularity of house music and the need to find ways of producing content that was popular in the clubs and distributing it to the population at large. Kwaito was distinguished by its slower rhythm or, as Bowes (2015) argues, its "sluggish" pace, "more suited to beer than E [Ecstasy]," pitched as those in the townships would be likely to hear it in the shebeens, beer halls, and local clubs. As several observers (Bowes 2015; Steingo 2008b) note, these early cassettes blurred the lines between cover versions and something slightly new, in keeping with much of the modern Afrodiasporic music-production ethos (perhaps most prominently displayed in hip-hop) where the lines between consumption and production are blurred. According to Tim White, a producer of early house and kwaito tracks who worked with several prominent kwaito pioneers, including Arthur Mafokate, "We were taking the big international house tracks, referencing them and doing our own slowed down versions of them: slightly Africanized, ripping a bassline here and there, so they were kinda cover versions but originals at the same time.... During that whole process—'OK, let's do it like this,' maybe adding a local snare—they were kind of getting a natural local influence. We were putting them out on cassette and selling them like fucking hotcakes out of the boots of our cars to various distributors around the country" (Bowes 2015).

As Steingo discusses, these early forays were often simply known in local parlance as "international" music (2008b, 2016). At what point "international" became kwaito is up for debate, but as cover versions began to morph into a more distinct sound created by both shifts in sonic texture described above and increasing lyrical vernacularization, kwaito began to replace the cover versions as the dominant form (Bowes 2015). Arthur Mafokate was instrumental in encouraging fans, fellow musicians, and media to adopt the umbrella term "kwaito" for this new music, which, well into the late 1990s and early 2000s, was known by a variety of names, including s'ghubu, s'waito, d'gong, and guz, with the latter namesake being the title of the TKZee Family's enormously popular album released in 2000 entitled *Guz 2001*. In the South African case, much of what was originally called "international" was mediated by a variety of factors, including circulating global capital, white producers and media commentators, and demands for soundscapes to accompany mass transit. I also ponder the perhaps less-acknowledged role of Afrodiasporic connections in facilitating the popularity of this new genre as a process of re-memory, a way to speak to the intersectional practices of embodiment that emerge in kwaito.

Lastly, the song "Nkosi" is itself part of a process of cultural exchange that reveals the song to be Afrodiasporic in content. Writing about the original composition by Enoch Santoga, anthropologists Bennetta Jules-Rosette and David Coplan (2004) argue that "Nkosi" must be considered a multivalent composition whose meaning shifts depending on performance context. Connecting the composition of "Nkosi" to the visit of Orpheus McAdoo's Virginia Jubilee Singers and the rise of the African Methodist Episcopal Church among Black South African elites, Jules-Rosette and Coplan demonstrate the numerous Afrodiasporic influences that permeated the composition of "Nkosi": "Enoch Santoga's compositions, including 'Nkosi Sikelel' iAfrika' were products of this syncretic cultural combination in a time of changing social consciousness and increasing cosmopolitanism among South Africa's Black elite" (2004, 350).

Jules-Rosette and Coplan remark as well on how migrant laborers from various African countries helped to circulate the popularity of these syncretic South African musical, religious, and political practices, introducing "Nkosi" into a number of new cultural contexts and spaces across Africa. As a result, "Nkosi" has served as the national anthem of five African countries and has been connected with African liberation movements throughout the twentieth century. Today, "Nkosi" serves as the national anthem of three African countries. Thus, while the song is intimately connected with the South African liberation struggle, it would be a mistake to view its performance as solely bounded by the national. In fact, its adoption as part of the South African national anthem implicitly links South African nationalism to global Black liberation struggles and to historically situated forms of Afrodiasporic circulation. When the young, erotically charged Black bodies of post-apartheid South Africa represented by Boom Shaka perform "Nkosi," they are already performing the nation diasporically. However, it is precisely the inability to contain the kwaito body within the nation that makes it a source of discomfort and anxiety for those who criticize kwaito adherents. Thus, the performance of Boom Shaka (particularly Mathosa and Seete) is an example of gendered and sexual nature of diasporic (mis)recognition. The musicians become less South African because their performance of gender and sexuality is said to contravene social mores. However, when reconsidered, their sartorial and kinesthetic borrowings participated in forms of Black (women's) exchange within Afrodiasporic Space that continue long histories of Afrodiasporic encounter and exchange. Unfortunately, rather than embrace this form of reconfigured subjectivity as being consonant with the very freedoms promised by the liberation movement, Boom Shaka and the kwaito bodies they inhabit were misrecognized as inappropriate with respect to gender performance, sexuality, and nation.

Freedom Remastered

Boom Shaka's performance of "Nkosi" suggests a reconceptualization of the category of freedom away from political and popular notions developed in post-apartheid South Africa. As Lindiwe Mazibuko, a former parliamentary leader for the opposition Democratic Alliance Party, argues, "Often we interpret our freedom in South Africa as an event. One which took place in 1994 and need never be revisited. True freedom, however, does not begin and end with the ballot box" (2011). Mazibuko highlights how, conceptually, freedom has been understood as rooted in the representational politics of the nation-state and control of the governmental apparatus, as a matter of being and not a matter of becoming. Therefore, according to one interpretation, South Africa has achieved freedom insofar as Blacks dominate the political process and control the government. Conversely, there are those who argue that freedom has yet to be realized since Blacks have little control over the economy of the country and whites still own a disproportionate share of the wealth.

Under both visions of freedom in post-apartheid South Africa (the neoliberal and the neo-Marxist), the state and its policies remain at the center of the analysis often to the exclusion of public culture. Furthermore, the nation remains both implicitly and explicitly heteropatriarchal. As such, the body remains an unexamined site for understanding and negotiating the disconnect between political control of the state and control of the economy. Cultural studies is dismissed altogether or analyzed simply as a superstructure to the economic and political base. For kwaito bodies, freedom means something potentially different from what was conceptualized either by the liberation movement led by the African National Congress (ANC), or by the movement's critics on both the left and the right.

Thus, any consideration of Boom Shaka's performance of "Nkosi" must grapple with competing definitions of what freedom is and what it allows one to do. As the members of Boom Shaka seductively and erotically performed their version of "Nkosi," they placed into crisis competing societal definitions of freedom. What did freedom enable the members of Boom Shaka to do relative to the wishes and desires of others (see O. Patterson 1991)? For kwaito bodies, freedom is often positioned firmly in this notion of personal freedom, given the lack of power they have to exercise freedom regardless of the wishes of others (particularly of their parents, but also of the state). This lack of power leads to a subsequent lack of interest in explicit participation in forms of civic engagement if we understand it to be limited to traditional forms of politics such as party membership, voting, and community organizing. This is not to say that the performances of kwaito are not "political"; rather it is to insist on the placement of the realm of the political outside of its familiar boundaries of electoral politics and into forms of body politics—perhaps remastering politics alongside conceptualizations of freedom. Rather than conceptualizing freedom as making the neat connections of the Rainbow Nation, Boom Shaka pushed for a recognition of freedom embedded in the messy practices of Afrodiasporic connections. These messy practices of Black diaspora open space to push back against gendered and sexual norms entrenched within the conceptualizations of the "New" South Africa. Similar to Black queers in Cuba studied by Jafari Allen (2011), Boom Shaka, and by extension kwaito bodies, uses practices of subject formation to contest limiting conceptualizations of post-apartheid Black youth. The musicians pursue a "larger freedom," one that is situated in practices of remastery—the shifts of subjectivity and space making I delineate here—in order to challenge statist and static forms of freedom.

Conclusion

Who then controls kwaito bodies? The contestation over Boom Shaka's performance was certainly about the ways in which the notion of personal freedom contrasted with ideas about freedom in relation to others. That is how the right to do what one wants is constrained by the rights of others to restrict your freedoms, particularly when such restrictions are ostensibly for the good of society. Boom Shaka insisted that the song "Nkosi" gave them the freedom to enact their erotic performance—after all, the liberation struggle was at least partially about the right to personal freedom. Speaking at the Generation Y keynote hosted by the YFM radio station, Mazibuko (2011) had this to say about what freedom means to the Y generation: "The attainment of substantive freedom can be measured by the extent to which South Africans are able to exercise choice in their daily lives unencumbered by the baggage of the past or the circumstances of their birth."

> I wanna be free from these chains that are binding me
> I wanna be free from these chains that are hurting me
> Somehow, somewhere, someday, somewhere, I'll find the place that true love lives
> I'll be free from this pain, free from the rain, free from these chains that are binding me. (Boom Shaka, "Free," 1998)

In their song "Free," Boom Shaka expresses the same sentiment. Backed by a pulsating house beat, the song is one of the few popular Boom Shaka hits that are missing the voices of the male members. Instead, the vocals of Lebo Mathosa, backed by Thembi Seete, dominate. But if the members want to be free from the chains that are hurting them, what represents the shackles that they need to be freed from and how might they go about creating this freedom? The answers to these questions lie in the divergent and unsettled responses to their performance of "Nkosi." Those who argued that solemnity and "respect" were central to the performance of "Nkosi" as "the national anthem" wished to enact a form of power over the public performances of young Black bodies. In the end, Boom Shaka performed a version of freedom despite the criticism. For the musicians, the song "represent[ed] their freedom to sing whatever songs they choose in a liberated nation and an acknowledgement of those who have fought for that freedom" (Blignaut 1998). As Junior Sokhela remarked, "It's [the apparent controversy over the song and the performance] a little bit of a misunderstanding. We are not dissing anything, this is our own version, one for the young people. . . . Our parents know the lyrics to that song but a lot of kids don't, even though they stand at school and hear it sung every morning. Young people's reaction to our version of the song has been incredible, they loved it. And this way they'll learn the lyrics too" (Blignaut 1998). Sokhela added, "Our aim was to sing the praises of all of the legends who sacrificed much in the struggle for justice and equality" (Pokwana 1998). While these seem like wholly uncontroversial aspirations, the controversy was situated in the erotic manner of the performance and the use of the house-influenced kwaito beat, which connected this version to young, Black, poor, and working-class township-based bodies and the criminality of tsotsi culture. Shifting the rhythms from hymnlike and solemn to celebratory and danceable also represented a chaotic noise that leapt past celebration and moved into the territory of anticipation and hence critique. Now that freedom is achieved, is it only to be celebrated or could we also ask what we can do with this freedom? In testing the limits of what kind of freedom could be performed, Boom Shaka both implicitly and explicitly critiqued the conservatism and (neo)liberalism of the post-apartheid contract. As a result, the song and the performance departed from a politics of bourgeois respectability. What kinds of new imaginaries of the nation get created when "Nkosi" is performed against convention by young, Black, erotically charged bodies? Which version of freedom is ultimately envisioned? And how is the nation imaginatively reconceptualized through these kinds of performances enacted by kwaito bodies? These questions and the ambiguity of their resolution are central to understanding the controversy. Boom Shaka felt empowered by "Nkosi" and its legacy to insist upon the erotic dimension of political struggle instead of simply presenting an image of Black suffering. Rewriting the presentation of the Black body, refusing its genealogy of stoic sacrifice and solemnity, Boom Shaka reminded the post-apartheid nation that the struggle also included the erotic pleasure made possible through sharing a common goal across difference. Because this aspect of community building is considered feminine, it is, as Lorde (1984) reminds us, denied or suppressed. In the "Nkosi" performance, Boom Shaka brings into the public sphere that which is unacknowledged and, in the process, remasters freedom.

As Boom Shaka prepared for its performance at the 1998 SAMAs, the group surely could not have anticipated the vigorous debate that the rendition of "Nkosi" would cause. Prior to the performance, the group had actually consulted with the Department of Arts and Culture and received its blessing (Pokwana 1998). To thwart criticism that Boom Shaka was using a symbol of national struggle for commercial gain, the group members agreed to donate all proceeds from the single to various charities (Blignaut 1998). The criticism that Boom Shaka might make any profit from the song seems overblown in an age of neoliberal capital. Certainly, political figures within the ANC consistently put the apartheid struggle to use regularly to enrich themselves. Lost in the debates about the controversial performance was the fact that Boom Shaka was exercising a version of freedom in order to create material gains personally, a template for which any of the politically connected Black bourgeois class surely could not fault. Prior to recording the album *Words of Wisdom*, from which "Nkosi" was the lead single, Boom Shaka signed an unprecedented record deal for young musicians performing kwaito that gave the group members

75 percent control over recorded masters and 100 percent copyright over any created material (Blignaut 1998). Boom Shaka became the first kwaito musicians to have such control over their music, and one of only two Black musical artists at that time in South Africa (the gospel singer Rebecca Malope being the other) to have a deal that situated so much economic control in their own hands. In this sense, these kwaito bodies rewrote the rules of labor in the music industry and exercised their newfound economic possibilities for Black youth. If nothing else, Boom Shaka's performance of "Nkosi" was an act of simultaneous personal and economic freedom.

The occasion of Boom Shaka's performance of "Nkosi Sikelel' iAfrika" has been used to introduce the guiding analytics of this study. First, any study of kwaito can be enhanced by an analysis of embodiment, centering on the performance of kwaito bodies as both musicians and fans. Kwaito bodies challenge the heteropatriarchal construction of the post-apartheid nation. Black feminist and Black queer theories are central to uncovering the ways in which kwaito bodies serve to disturb the post-apartheid fantasy of the active Black male heteropatriarch as the central component of a postliberation society. Afrodiasporic Space is invoked in order to illuminate the ways in which the construction of the South African "nation" post-apartheid is premised on forms of exclusion that are central to any nationalist project. Kwaito bodies reveal both implicitly and explicitly the fallacy of the nationalist concept and the violence that such a concept visits on the most vulnerable members of the South African polity. Furthermore, the mobilization of Afrodiasporic cultures in kwaito is part of a long history through which young Black people have contested constricted definitions of Blackness in South Africa (Livermon 2012; Pietilä 2013). As a precaution, the diaspora analytic is not meant to be a totalizing narrative that obliterates alternative understandings of the post-apartheid South African nation. Instead, diaspora as an analytic becomes one of the ways in which to ponder the exclusions of the post-apartheid nation. Diaspora does not and cannot account for all forms of cultural contact and exchange in post-apartheid South Africa. However, my observations of kwaito music and culture and the performative bodies that inhabit Afrodiasporic Space are shaped by the active engagement between myself as researcher and my Black South African interlocutors. Ultimately, the question of how freedom should be lived and enacted is what is being struggled over as cultural commentators, Black youth, and senior political officials debate the meaning of Boom Shaka's performance. As we shall see, performances enacted by kwaito bodies, both onstage and off, will be an important arbiter in the struggle over the meaning of freedom for post-apartheid South Africa.

2

Jozi Nights

THE POST-APARTHEID CITY,
ENCOUNTER, AND MOBILITY

In the early 2000s, Capital had become the hottest new "it" club among Johannesburg for socialites, the Black bourgeoisie, and the cognoscenti—a different crowd from the kwaito fans I had come to identify with. On Friday and Saturday nights, I would drive by the bar-lounge in Rosebank and witness the lines of patrons waiting to get inside. I wondered what was so special about the space, its crowd, or the DJs who filled the night with music. In the previous months, friends of mine had waited quite some time to be admitted, only to be turned away at the door: they figured the club had an arbitrary or even discriminatory admissions policy. One night, a friend of mine—a club regular—invited me to meet him there, to dance and explore. But I could not just walk inside. I was to text him when I arrived with my friends and he would, in his own words, "make sure I got in." The door was ruled by two funky, well-dressed women who looked patrons up and down to make snap decisions about who was worthy of entrance. One gatekeeper was Black; the other was white: all the same, I noticed that many of the Black patrons were turned away.

Those who went to clubs and parties knew that, like the rigorous, but rarely publicly spoken, gender ratios at some clubs, many of Johannesburg's "mixed" clubs (those that catered to both white and Black South Africans) wanted to maintain appropriate racial ratios. Club promoters designed these unspoken racial ratios to create the sense of mixed-race spatiality that allowed higher-spending, young white South African clientele to perform the neoliberal multiculturalism of the era without making them uncomfortable. Importantly, this kind of mixed-race spatiality was classed, given both the locations of these clubs, their door policies, and the cost of buying food and drink. Allowing the club to become "too white" would lessen its appeal to the Black clientele that gave the space its coolness. However, allowing the club to become "too Black"—especially admitting the "wrong type" of Black person—would disrupt the air of exclusivity and the class performances that were central to the brand. While in many ways these upmarket, multiracial, middle-class club spaces traded on the cachet of Black working-class and Black township culture, they shifted away from playing those audiences' beloved kwaito music and dedicated much of their playlist to house music (both local and international) and international (mostly US) hip-hop.

By the time I had parked my car, the line was snaking down Tyrwhitt Avenue, away from the entrance to the club, reaching practically to the next block. I overheard patrons both complaining about the line and excitedly speculating about if or when they would be admitted. In the background, I heard the kind of European house-techno fusion music that would become contemporary global electronic and dance music in the vein of Calvin Harris. As the sounds of middle-class cosmopolitanism played in the background, we passed the entrance to see the cool crowd lounging around the tables inside and waited for our turn. After months of passing by the club with the sense that I would never be admitted, there I was, in line with everyone else. I nervously looked forward to gaining entry, knowing that many in the line before and after me would not. And I was ashamed to have fallen for the hype. Even so, on that Saturday night I was anxious about knowing someone in the "in" crowd who could ensure my admission. It is quite interesting how exclusion can produce desire, the need for belonging, and a feeling, no matter how fleeting, of triumph even among those such as myself who are trained to know better.

As soon as we joined the line, I texted my friend, a well-known stylist for the celebrity set. He had a crush on my working-class friend from a West Rand township and insisted that I bring him along. My friend accepted and brought another fellow from his township for company. I realized that I was trading my friend's presence for the price of admission, but my entourage was eager to gain admission to Capital. One of their township cohort had experienced the club and bragged about it constantly, talking about the celebrities he had seen, and the "high-class" atmosphere and the selection of music, which was not typical township shebeen fare. Within minutes, my stylist friend appeared, pulled us out of the line, shared a laugh or two with the young women at the door, and mentioned that we were the "guests" he was expecting. Like that, we were behind the ropes and into the lounge. The experience itself was so unremarkable that I wondered if the process of gaining admission would be the most eventful portion of the evening. I found the club, its mixed-race, middle-class clientele, and the music (which on this evening veered toward European techno) to be bland and nondescript. I was left wondering: "What was the big deal?" And yet I understood intimately why people wanted to be there and the kinds of possibility the space presented. I experienced Capital as a microcosm of the shifts in the Johannesburg nightlife scene: a glimpse of what a mixed racial and class social scene could look like, even if that mixture was predicated upon implicit and explicit erasures and exclusions.

By the early 2000s, the party and club scene of Johannesburg had shifted dramatically from my mid-1990s forays into Yeoville. Kwaito and the attendant club cultures it spawned had arrived in different spaces. The kwaito universe had in the intervening years become much more diffuse, moving from a series of bohemian and queer Johannesburg club spaces and informal shebeens and taverns in townships of Soweto and Alexandra. Occasional club and party spaces still appeared in Yeoville and Hillbrow, but now they were marked as working class and poor: no longer considered cosmopolitan, they were defined by the understanding that these places were unpretentious—anything goes. In fact, without quite realizing it, I noticed that I had nearly stopped partying in Yeoville altogether, and Hillbrow's trendsetting status was starting to die out.

Club spaces increasingly centered on notions of exclusivity, meant to appeal to South Africa's emerging communities of aspiration. Clubbing in Johannesburg began to mimic, but not necessarily replicate, the spatial geographies of exclusion that defined Johannesburg. On the one hand were more informal spaces like shebeens and inner-city nightclubs that featured inexpensive drinks and no cover charges. These clubbing spaces were to be found primarily in townships and inner-city areas of Johannesburg, including the Central Business District (CBD), Braamfontein, Newtown, and the older original incubators of kwaito club culture, Yeoville and Hillbrow. On the other hand were a proliferating number of club spaces located in Rosebank, Sandton, and Randburg, as well as Soweto that were defined by cover charges and/or expensive drink lists.

This chapter looks at how Black youth use practices of leisure and partying to remake urban space in Johannesburg. My argument is that, while nightlife often reinforces the spatial divisions of the political economy, it also provides opportunities for the traversal of these same divisions. As was demonstrated above in the scene from Capital, exclusionary space relies on and needs the very people it is supposed to "exclude" in order to function. Hence, my focus is less on demarcating these exclusions and more on highlighting the creative and engaged ways young Black South Africans use nightlife to express models of post-apartheid freedom, which may be reactionary, subversive, revolutionary, or imaginative.

To document and engage these micro-practices of remastered freedom, I take the reader through several nights of partying that my friends and I experienced in Johannesburg. How does kwaito help to create an alternative geography of the city, and what does that alternative geography reveal? Kwaito bodies conceptualize the city at both the macro and micro levels through the practices associated with partying. These spaces create possibilities for remastering narratives of encounter in post-apartheid South Africa. To remaster encounter, I engage Paul Gilroy's concept of conviviality. For Gilroy, conviviality speaks to the processes of "cohabitation and interaction" that mark contemporary social life (2004, xv). Conviviality "does not describe the absence of racism or the triumph of tolerance" (xv). Rather, it captures those very moments when the rigid conceptualization of race, class, gender, and sexuality prove to have moments of instability. Drawing from the concept of *motswako*, a Setswana term meaning "mixture," the soundscapes produced by kwaito bodies in space require the deployment of an alternative conceptualization of encounter in the South African present. Lastly, I provide an analysis of my own performing body as a co-performative witness (Conquergood 2002) who experiences these moments of pleasure and possibility alongside, and in partnership with, my communities of research. By looking at how Black people remobilize and reorganize space, the city is revealed as a place where the hierarchies both cohere and collapse under the weight of traversing kwaito bodies.

Johannesburg is often spoken of as a city of despair marked by enormous inequality, horrific violence, and the triumph of neoliberal capital. In "From Jo'burg to Jozi," the introduction to his coedited book *From Jo'burg to Jozi: Stories about Africa's Infamous City* (2002), journalist Adam Roberts suggests that popular nicknames for the city of Johannesburg point to important shifts in perspective. For Roberts, these shifts in perspective are important for reclaiming the city and enacting a more egalitarian understanding of its social fabric. He writes, "Something, perhaps the lure of money, excitement, work, bright lights and opportunity, something continues to draw thousands of newcomers to the city every day. These people see hope in the city, not despair. They also have a different name for Jo'burg, one that sheds all connection with the old, white mining city and instead promotes its youthful, African, edgy reputation. They call the city Jozi" (6). Roberts argues that the nickname Jozi signifies not just a semantic change but an analytical reworking that centers Black South Africans and reconfigures Johannesburg as a global African city.

Anne-Maria Makhulu argues for the political significance of Johannesburg youth:

> This [the lack of Black people in central Cape Town] perhaps distinguishes Cape Town from other cities and, in particular, Johannesburg, where a growing Black middle-class works and plays. But Johannesburg is also, cliché though it may be, genuinely an "African" city in a way that Cape Town isn't. Whatever people's income brackets and employment status, one rarely gets the sense that Black residents of Johannesburg feel somehow unwelcome downtown or even in the swank Nelson Mandela Square in Sandton City Shopping Center. Johannesburg is abuzz with hip young Black people of every class and income bracket. (2015, xiii)

In this brief sketch of the city, I draw from both Roberts and Makhulu in understanding the shifting cultural and material geographies of Jozi. As a shorthand, the city can be effectively divided into three major nodes of circulation that are codependent: the northern suburbs, the inner city, and the townships. Each of these three spaces has undergone significant changes post-apartheid as Black South Africans have reconfigured these areas through the materiality of their presence: their strategic choices in housing, and their critical creativity in the realm of leisure.

The northern suburbs of Johannesburg are collectively the wealthiest part of the continent of Africa, and as such they retain a colonialized, racialized hierarchy in the spatiality of the city. Created in the 1960s out of what was once farmland, the northern suburbs were an illusive façade of late apartheid South Africa. Like much of urban South Africa, these suburbs were as much ideological as they were a result of shifting political economies. They represented the supposed efficacy of apartheid economic and social policy. They were made possible by the explosive growth of the post–World War II South African economy and the need to accommodate a growing, more prosperous, white minority. Located north of the traditional city center and the original white, wealthy enclaves of Parktown and Westcliff, these growing suburban areas were significantly connected with "new money" elites who made their money not through the traditional route of the extractive economy, but through services, information, and finance that were growing with the globalizing economy. Starting in the 1980s, these areas began to see extensive redevelopment as businesses and the outposts of global capital began to abandon the city center. As a result, the northern suburbs (Sandton in particular) became the site of residential and commercial wealth as a new downtown was created; this centered its development in Sandton City. Concurrently, upmarket shopping and leisure spaces that catered to the new and old bourgeoisie proliferated. Malls (a particular American suburban spatiality) were key to these new leisure and consumptive spaces because they restricted entrance to those who could afford to consume. Key to understanding these spaces is that Blacks, no matter how wealthy, were restricted from engaging with these areas as consumers or residents. They could only exist as laborers—in most cases, the domestic variety where men served as gardeners or manual laborers and women served as maids and cooks. This created a unique spatiality whereby Black women (and less often men) resided in these suburban spaces behind the homes of the white families they toiled for, separated from their children and family members who resided either in townships or more often in the rural homelands. It was considered particularly liberal and enlightened to allow the family of the help to reside on the property. More liberal still were the few families who supported the children of their domestic workers by socializing them into middle-class schools, thus guaranteeing that these children of servants would be among the first to enter into the Black middle class post-apartheid. While the terms of inclusion were strictly unequal, what was clear was that even when the northern suburbs lacked Black residents, there were Black residents (see figure 2.1).

In the post-apartheid period, these areas have undergone change in their racial, if not their class, characteristics. The entrepreneur and urban developer Richard Maponya, with the help of white friends, was able to live in Sandton clandestinely, in violation of the Group Areas Act that was the main policy that dominated urban spatial provision during apartheid. The act allocated central, well-developed areas to whites, while leaving peripheral, less developed areas to Blacks. By the early 1990s, with the easing of most apartheid laws, these areas began to experience various forms of racial integration. They became important signifiers of Black middle-class status, both for the amenities that they provided residents (consistent water and electricity, paved wide roads, security, and proximity to economic nodes and high-ranking schools) and for the sense that Black South Africans were reclaiming spaces of exclusion. Rosebank, the neighborhood at the center of the opening vignette, is perhaps one

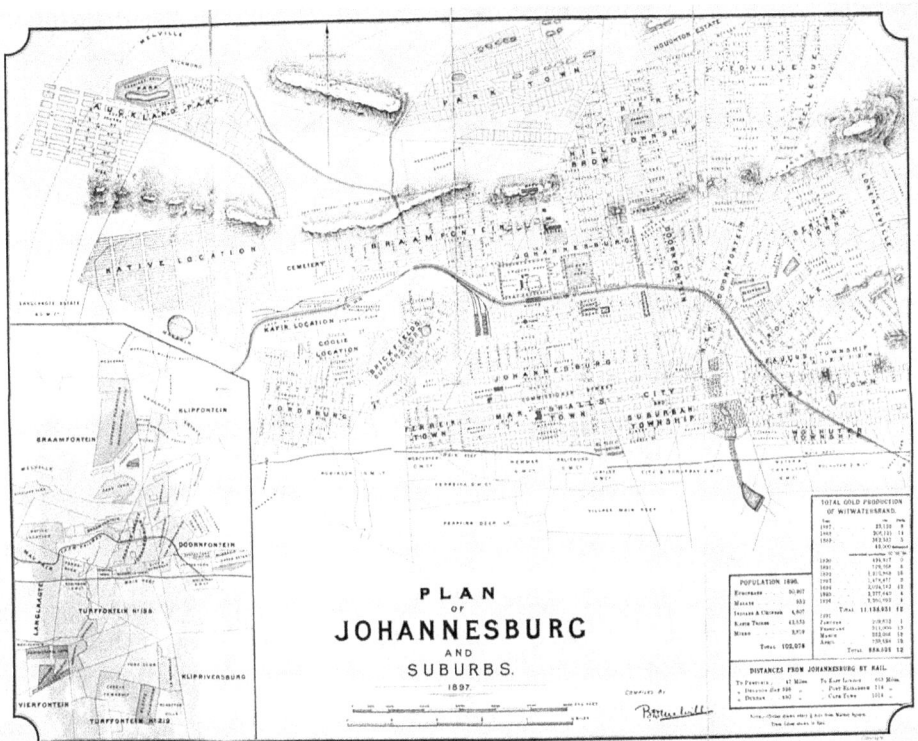

FIGURE 2.1. 1897 plans of Johannesburg and surrounding suburbs, which would guide city planning throughout the 1960s. Source: Library, University of the Witwatersrand, Johannesburg, South Africa.

of the starkest examples of population shifts within middle-class suburbs in the area. While Rosebank is a small (0.82 km²), mixed-use neighborhood with significant commercial development centering on shopping and leisure activities, a population of more than one thousand people live there primarily in apartments and condos. During the 1990s, there were few Black people living in Rosebank. Today, 46 percent of the population is Black.

The effect of living in these spaces is both material and symbolic. Recent tweets by young, middle-class Black South Africans reveal this duality. Luyolo Mphithi tweeted this upon closing on his first place: "Bought my first property. Ugogo [grandmother] was stunned. Cause she wasn't coming to this area as a domestic worker anymore" (Mphithi 2017). The basis of purchasing the home in this area where older Black women serve primarily as domestic workers is embedded in the materiality that allows him to afford the residence, one that is heavily structured by race and gender. And yet the symbolic importance of his grandmother occupying the space as an invited guest of her grandson (rather than a domestic worker) is key to the symbolic imaginary of these spaces. This is particularly so, given that many Black middle-class workers are unable to count on generational wealth to pay the down payments necessary to afford housing in these areas, must dedicate substantial percentages of their income toward supporting family members, or are not well compensated enough by their companies to afford to purchase in these areas. Nevertheless, the racial composition of many of the areas in Sandton has become increasingly Black, and according to the 2011 census, 35 percent of Sandton's residents are Black. During the early 2000s, leisure spaces were demarcated by race and class. While a number of mixed-race venues developed in these northern suburbs (and Capital, discussed at the beginning of the chapter, is one example), many of the spaces used music, door policies, cultural influencers (such as DJs and celebrities), dress codes, and nascent social media platforms to frame their clientele more exclusively by race and class.

While there is no actual demarcation of areas of Johannesburg as inner-city, the term has come to be used to define the traditional CBD and a number of suburbs that surround the CBD, particularly those suburbs that have shifted dramatically in both racial and class composition. These suburbs include Hillbrow, Berea, and Yeoville, as well as the student-centered suburb of Braamfontein and the arts-oriented Newtown. The unique tensions of the simultaneous processes of investment and divestment that texture these neighborhoods are particularly important to the lives of residents and to the cultural landscape of post-apartheid South Africa. In the 1990s, as the Group Areas Act was enforced less, Black South Africans began to move to inner-city areas, particularly to the bohemian and countercultural spaces of Yeoville and Hillbrow. Both the apartment complexes of the CBD and the adjacent areas of Hillbrow and Yeoville underwent dramatic change as these areas increasingly became sites of Black migration into the city. This migration was threefold, merging Blacks from local townships, Blacks from elsewhere in the country (both other urban areas as well as rural areas), and African migrants from the rest of the continent. Around the same time that these areas were becoming increasingly Black and increasingly African, there was (white) divestment from these areas. This divestment was in the form of commercial property as companies relocated to the northern suburbs (particularly Sandton) and residential disinvestment as buildings were abandoned.

Migrants to these inner-city neighborhoods might land in well-cared-for housing with full amenities, or they might occupy a building that had been abandoned by its white owners. As workers and homeowners passed through, regional demographics were in constant flux, marking them as liminal spaces, full of make-do and possibility. Middle-class students and their Black bourgeois counterparts would eventually move on as their incomes and statuses rose. Many of the new migrants, both from South Africa and from outside South Africa, would also hope to eventually move out of these areas even if their choices were more constrained by their lack of mobility in the post-apartheid economy. But the liminal vibrancy and possibility of these neighborhoods remained as migrants came and went, hoping for a better life than whatever they had left behind could promise.

This inner-city space represented a constantly shifting demographic intersection, and it contained the kinds of transitional subjectivities, community values, and life trajectories that emerged in the post-apartheid transition. It was peopled with newly mobilized citizens circulating between employment in the northern suburbs, to social activities in the townships, to living in the CBD: a series of belongings that would have been impossible in the entrenched boundaries of apartheid. Many of my friends moved from the townships, to the inner city, to the northern suburbs as income and resources improved. In this way the inner city often negotiated the stark geographic contrast between the leafy northern suburbs and the townships/rural areas/other African nations. From the early 1990s to the early 2000s these neighborhoods, contrary to the vision of the Group Areas Act that imagined these spaces as white-only zones, become almost exclusively Black. While initially a space where the Black lower middle class, students, and Black new middle classes congregated to be closer to city amenities, the inner-city areas, particularly in abandoned buildings that had been taken over by residents, increasingly became a place for the working poor and their families, who could not afford other forms of accommodation. That the material circumstances of inner-city residents shifted from block to block accounts for the cultural practices of imagination and self-writing that emerged within them. Importantly, these areas nurtured many of the first kwaito club scenes during the early 1990s. Because acquiring abandoned postindustrial space was inexpensive, these inner-city areas often hosted the underground of South Africa's music and art scene. You could never be sure the venue was legal as you slipped past security and listened to the latest poetry reading, deep house session, kwaito groove, hip-hop competition, or jazz band, but it did not seem to make a difference. What was clear was that unburdened from some of the costs associated with venues elsewhere in the city, the inner city had once again begun to nurture the creative underground of the city. By the early 2000s, the city of Johannesburg (through the Johannesburg Development

Agency) began to turn its attention back to the inner city of Johannesburg, with a mind to redevelop these areas. Two processes were now in motion. The first was the seizure and recapture of illegally occupied buildings with an eye toward redevelopment. Authorities perceived these spaces as harboring criminals, drug dealers, prostitutes, and purveyors of related urban ills. The second process was the designation of areas as redevelopment districts, with numerous real estate incentives typically available for white, wealthy investors to refurbish buildings and move into these areas as an alternative class of residents. These twin processes have probably been most prominent in the CBD areas of Jeppestown, Newtown, and Braamfontein; Hillbrow and Yeoville have seen the mass evictions without the high measure of redevelopment. These complex and uneven forces of gentrification have had mixed results. In some areas, there have been concerted efforts to offer affordable housing (typically still far more expensive than the previous inhabitants can afford). In Jeppestown, residents from the nongentrified areas pass through these spaces in such a way that local businesses such as hair salons, corner stores, and bars have shifted to accommodate them while high-end cafés, bars, and restaurants have not. Club and party spaces, somewhere in between, have become an important way in which young Black South Africans would occupy and make space outside the confines of state and corporate mandated redevelopment.

As Achille Mbembe writes, "The township both is and is not urban, it is proximate to the city while at its margins" (Mbembe et al. 2008, 239). His important observation frames understandings of township space. Township spaces are contradictory, capable of containing disparate truths and practices of quotidian life. In the post-apartheid moment, the government reconfigured city space so that the townships (mainly Alexandra and Soweto) have been integrated into greater Johannesburg. Yet the sense of being simultaneously *of* the city and distinctly *apart* from it remains. Significant upgrades in infrastructure and amenities have occurred throughout South Africa's townships, yet the neoliberal nature of privatized services has meant that poor families often struggle to access these amenities. Township space is highly variegated, ranging from the upper middle-class and wealthy enclaves of Diepkloof Extension to informal settlements nestled next to Freedom Square in Kliptown. The idea of the township landscape of poverty is complicated by the presence of Black middle-class areas and the capital they bring to bear.

Two seemingly disparate narratives concerning township space highlight the contradiction; these frame my approach to understanding the townships and post-apartheid urban policy. On the one hand, a new infrastructure and an influx of previously unavailable amenities altered the landscapes of the townships, remaking them into livable spaces for the majority of their inhabitants (even as these spaces are increasingly privatized and commodified). On the other hand, it is clear that most township residents must find and secure income outside of the townships. Hence, township life requires a sense of perpetual motion. As Nsizwa Dlamini puts it, "The township is also a space in motion. People are perpetually moving and commuting. This sense of always being in motion is captured in the language of township residents, their dress code, and their music—kwaito in particular" (Mbembe et al. 2008, 241). This sense of mobility, which animates the sonic practices of the township, is one captured by the concept of motswako. Spaces of leisure are the product of a series of calculated movements: embodied practices of self-making and community-making on the move. Young Black South Africans circulate into and out of the space of the township; this mobilization for opportunities and experiences also emplaces them in the club life of the city.

The Encounter

In his text *Times Square Red, Times Square Blue*, Samuel Delaney (1999) discusses the urban milieu as a space of encounter. Drawing from urban studies pioneer Jane Jacobs, Delaney describes encounter in urban settings as the opportunities created by urban institutions that enable interclass contact. While his particular interest is the space of queer men's encounter facilitated by the (il)llicit sex venues of pre-2000s Times Square, his idea that "life is at its most rewarding, productive, and pleasant when large numbers of people understand, appreciate and seek out interclass contact and communication" is at the center of my examination here (111). Nightlife spaces allow for precisely the kinds of encounters that destabilize the hierarchies of post-apartheid South Africa, hierarchies anchored in class, but also delineated by race, gender, and sexuality. Delaney deftly weaves his own experiences of cruising Times Square with his observations regarding the violent erasure of what he identifies as critical spaces of freedom. This negotiation between personal experience, the experience of others, and embedded social history forms a method of reading urban space. He centers his own performing body in relation to the city and its eco-movement with others: a performance ethnography by which the city is read and produced by the body-in-motion and in encounter. Embedded in the erotics of contact, these sites of ethnographic experience offer a rich method for understanding the relationship of the experimental, exuberant body to the city. To attend to these moments in my work, I borrow Delaney's language and refer to them as "encounters."

Kwaito bodies situate themselves within and through the space of the city as they negotiate the complicated contexts of post-apartheid life. These seams of negotiation, of self-articulation and lived expression, come into creative tension in this ethnography: they offer the conceptual frame through which I understand the ways that pleasure is experienced within spaces coproduced by young Black South Africans. The encounter heightens the sensory perception of the moment, in particular the pleasures of the body that are key to how space is experienced within the soundscape of kwaito. Thus, the encounter is a site of ethnographic analysis. The encounter serves as an ethnographic vignette that examines the textures of sonic and spatial coproduction. These moments of pleasure, and the erotics of remastered freedom they are capable of producing, are organized in terms of the overarching theme each encounter illuminates. Even so, each example lends itself to the frameworks and context of the others. Together, they compose a lexicon by which we understand how kwaito's performing bodies reflect, articulate, and act upon the conditions of post-apartheid. The theories that guide this ethnography spring from these encounters: a series of interplayed understandings and interpretations that dance with kwaito's multiple performative registers. In turn, each encounter illuminates a different aspect of kwaito bodies offering a way to read young Black bodies and their constitutive pleasures within narratives of power. In this chapter, three different encounters are examined that reveal Jozi as a city where practices of nightlife allow kwaito bodies to remaster urban space.

Encounter I: Mixing It Up, Mzansi Style

One possible outcome of encounter is that something new is produced, and often what is produced is described as a mixture. One critical process that has been important to understandings of the African Diaspora and has been increasingly investigated in South African cultural studies is the notion of mixture. In the post-apartheid moment, theories of culture have been rewritten away from regimes of purity. How does a consideration of Afrodiasporic Space in relation to kwaito create alternative imaginaries of encounter? Theorizing from the cultural practice of kwaito in relationship to the established Afrodiasporic analytic may yield alternative understandings of the idea of mixture in South Africa. Kwaito practitioners show that kwaito culture (here examined through nightlife) produces different processes of mixture that are distinct from the essentializing scripts concerned with racial mixing.

I discuss mixture as a possible outcome of encounter for two reasons. First, it is often the "mixed" and nonautochthonous nature of contemporary Black

South African youth culture that allows their engagements to be dismissed. With the exception of youth cultures used to rally political support (Steingo 2007), the cultural creativity of young Black South Africans is judged by elders and government alike as a dangerous devolution of "traditional" African cultural practices. While it may be an accepted axiom outside the African continent that Black cultures are a product of mixture, within Africa there is a prominent common sense that devalues practices that are framed as too divergent from supposed original sources. Second, the language of mixture (particularly in relation to musical style, sartorial practices, dance, and language) was prominent among kwaito musicians and fans. Critically, the conversation concerning mixture in the contexts of kwaito, and Black South African youth culture more generally, often elided discussions of racial mixture so prominent in these similar discourses in the West. To explore these ideas further, I introduce motswako to account for the rich ways in which locals conceive of cultural "mixture." The concept of motswako illuminates the shortcomings of racial binaries in understanding encounter, which will be used to critically reflect on processes of mixing outside the tendency to radically equate mixing with racial mixture. Returning to the opening vignette, a consideration of motswako would require us to think equally about the relations between differently classed Black people as much as (or perhaps more so) than the presence of whites and Blacks in the same space.

The dance floor radiated with the tunes of Brickz, the latest kwaito star coming from Soweto. In the mid-2000s, Brickz, with the help of his producer DJ Cleo, was bringing an old-school flavor back to kwaito and reviving a genre that had been in a bit of a lull. This meant a return to the slower pace (in beats per minute) of the mid-1990s; a re-centering of dancehall vocal styling and production amid the slowed-down, house backing track; and a vocal style that rested comfortably between Caribbean-inflected toasting and singing. Given Brickz's gospel background and the church training of much of South Africa's musical talent, DJ Cleo deftly weaves in electronic organ notes in many of Brickz's biggest hits including the one being played on this evening, "Sweety My Baby," where the electronic organ creates a dissonant and haunting melody as well as an alternative rhythmic timbre to the underlying house beat.

Glancing at the well-lit space, I noticed bodies writhing to the rhythm. Women danced with women, men with men, queers with straights, wealthy with poor. My partner was engaged in a conversation with someone sitting next to me. They were speaking in Sotho, a language I was starting to understand and speak. When it became clear to me that I had somehow become the focus of the conversation, I turned my head away from the dance floor and to my immediate surroundings. From what I could gather, the young man was interested in me. Never particularly jealous but eager to clarify things, I heard my partner say, "Ke motho oa ka," literally meaning "He is my person." In this context, he meant, "He's my boyfriend." "Wow," the guy said to my partner, "he's nice." I was integrated, desired, engaged in co-movement.

My partner exited the conversation, and at that point his acquaintance turned to me. "U tsoa kae?" (Where are you from?), he asked in Sotho. I hesitated, since, as I have previously discussed, most people did not generally read me as American in these contexts, and I find that answering truthfully can sometimes be met with incredulity or even hostility. Some people think that my claiming of American nationality (something I do with a measure of ambiguity) is tantamount to putting on airs. I am always aware of this and am often reluctant to reveal my nationality for these reasons. Sometimes I will simply say I am Nigerian, since this is what people expect, and it makes small talk easier. On other occasions, I will playfully claim to be from South Africa and create a complicated story about being raised in exile. These ethnic-national performances, which are simultaneously racialized, were my way of negotiating Afrodiasporic Space: of testing the context into which I had inserted my body, and of finding ways to perform selfhood within it (see Diawara 2003). This occasion I decided to be truthful, but I deliberately chose to respond in Zulu since I was more fluent in that language. "Ngiphuma eMelika," I answered. He was in disbelief. "No, really," he said. "U tsoa kae?" Again, I insisted on my American nationality. This time in Sotho: "Ke tsoa America." He said, "No, it can't be." He switched to Zulu, "Ngicabanga ukuthi uphuma eZimbabwe," meaning, "I think you're from Zimbabwe." At this point, I turned to my partner for confirmation. "Will you please tell him where I am from," I said with a mixture of humor and exasperation.

I drifted out of the conversation, but could overhear my partner and his acquaintance discussing my national origin in Sotho. Finally, the acquaintance asked in Zulu, "Igama lakho ngubani?" (What is your name?) I answered back, "Ke Xavier," in Sotho. Although he was still not convinced of my nationality, he began to speak in English with me, asking how I found South Africa and what were the differences between the United States and South Africa. At this point, I had better things to do. Sindi, one of the best dancers at the party, beckoned me to the dance floor, and there was no way I was going to pass up the opportunity to dance with her. As Brickz was replaced with Boom Shaka, we began to grind to the beat, singing the Xhosa lyrics, "Ndinaluthando nga phakhathi" (I have a deeper love inside).

Coming home from this house party in Soweto, I remarked at just how vibrant kwaito culture is and how it is able to account for a dizzying array of difference, yet simultaneously create a space of affinity. The mass of sweaty bodies grinding on the dance floor is an assemblage of complex, uniquely positioned people united in their pursuit of pleasure around this Afrodiasporic music. Mixed on the dance floor are different bodies. As mentioned previously, the music is also mixed. This mixture occurs within its composition of different Afrodiasporic elements (local and international) woven together. During the party itself, Black musics come together from a host of trajectories: dancehall, American hip-hop, house, and R&B are intertwined like the dancing bodies themselves. We easily switch between speaking in English (usually while addressing me), local slang, urban vernaculars, Nguni, and Sotho-Tswana: here, in the shared context of musical experience, we elide boundaries, designations, and barriers. Simultaneously, something new is happening. While the music is being pumped to its loudest volume, while different kinds of people grind against one another, what is being created? For a moment, kwaito produces a unique series of performativities and interactions that are unique to the space, and to the situation at hand. People of different nationalities, language and ethnic groups, genders, and sexual preferences come together in this space and make meaning across difference, creating affinity and moments of misrecognition for the hours we are there. These encounters produce mixtures for sure, but what kinds of mixtures are they?

There is a concept of racial hierarchization and racial mixing that is central to many theories of mixture. Racial hierarchization speaks to both the creative possibilities and the limitations of rereading South Africa as an Afrodiasporic Space of encounter. The idea of racial hierarchization limits the applicability of theories of mixing in South Africa borrowed from new world contexts. Racial hierarchization points to the notion of mixture as a process of dispossession, terror, and loss. While there can be no doubt that much of what shapes South African society and culture is predicated on a history of racial hierarchization, it is also true that much of the mixing that occurs in South Africa is not built on this premise. For instance, if we understand mixture to be produced out of a simple duality—a hierarchy that places white over Black—how do we account for the Black-Black intermixtures that pervade much of South African society? Afrodiasporic Space unfolds beyond the reach of the white-Black hierarchy, queering the master-servant and north-south binaries and undoing the work of white supremacy. In the process, the relations of Afrodiasporic Space forge opportunities for remastery through creating circuits of possibility that engage affinities among and between colonized populations. It is the product of south-south cultural interactions: the more complex tides that work on modernity's processes of cultural contact and exchange.

Returning to the party scene in Soweto, it would seem that the lens of racial hierarchization is inadequate or even inappropriate for approaching the thickness of this cultural encounter. The Xhosa-Zulu intermarriage that was a result of urbanization in Durban formed something new, an intermixture that dances with yet elides the clear-cut hierarchies of the plantation. The influence of reggae music and Rastafarianism in Black South African communities demonstrated the complexity of global Blackness as it articulates to diasporas, mobilities, and in-betweennesses of many kinds. Theories of encounter based on mixture reach their limits when they are unable to account for intermixtures that result from alternative interactions that formed beyond the stark Otherness produced by racial differentiation. While power, subjectivity, identity, and hierarchy certainly come into play in these spaces, the question of kwaito mixture cannot be read as solely contingent on the extreme hierarchizations (particularly of race) that typify slavery and colonization. Hierarchies of gender and sexuality, as well as of class, citizenship, and belonging, are omnipresent, for instance. The mixtures that occur through nightlife mark something more complex, more contingent, and, in the context of post-apartheid South Africa, more powerful.

The complex of Black positionalities (particularly in relationship to notions of humanity, modernity, and citizenship) that brings together Afrodiasporic Space mark these modes of encounter as fundamentally different from the differentiations and intermixtures that are produced from the systems of plantation slavery. Kwaito mixtures are not predicated on shades of difference between Black and white identities. Instead, mixtures occur within and between a spectrum of modes and experiences of Blackness, all facilitated by the textures of Afrodiasporic cultural style. The moments of entanglement that emerge through nighttime leisure are central to the formation of new Black identities in South Africa.

Jocelyne Guilbault describes the possibilities that arise from the forms of entanglement between the sociocultural and the sonic as "audible entanglements." Audible entanglements "*foresound* sites, moments, and modes of enunciation articulated through musical practices. . . . [A]udible entanglements assemble social relations, cultural expressions and political formations" (2005, 40–41, emphasis in original). Audible entanglements have the power to "render audible and visible specific constituencies, and imaginations of longing, belonging, and exclusion" (2005, 40). The "audible entanglements" of kwaito produce moments of affinity between and through ideas of Blackness that may not exist in any other space. This is an encounter that is not solely contingent on racial hierarchy and might best be accounted for through different terminologies. In this instance, notions of mixture must be considered beyond their histories of racial entanglements, racial hierarchies, and the fundamental African-European nexus that guides it.

Motswako is a concept that has emerged within some of the musical genres of post-apartheid South Africa: it marks the thick and improvisatory practices of mixture that make kwaito culture unique. Motswako refers not only to musical mixtures but also to the mixtures of cultures, languages, ideas, and space within contemporary South Africa. It also refers to the variety of possibilities and spaces of interaction through and across modes of difference. These differences are not necessarily racial; they are more likely to be of class, location, or nationality. Motswako is a useful reference to the process of encounter, the results of encounter (which we might call mixture), and the sentiment of encounter. Motswako has connotations of newness formed from contact, intimacy, and difference. In addition, it has come to define a musical style in post-apartheid South Africa. Motswako is one way to think about the specific types of encounters that define urban Black South Africa. The fact that these mixtures involve both situated and nonsituated positions accounts for the difficulty in sounding their contours. That is to say that these mixtures simultaneously represent both the intense intimacies of everyday lived experience and the forces of imagination and politics that animate Afrodiasporic Space.

The term "motswako" appeared most prominently during the television talk and lifestyle show of the same name. Featured on the South African Broadcasting Corporation's Channel 2 (SABC2), *Motswako* is targeted to middle-class women who speak Sotho-Tswana. Initially, the hosts of the show were three South African women (two Black and one white). The white host could not speak any of the Black African languages engaged on the show; as a result, the show's format was by necessity multilingual, in addition to being multicultural and multiracial. Even after a Black South African replaced the white cohost, the program continued to alternate between a Sotho-Tswana mélange and English. This required the target audience to have facility in both languages. There were even portions of the show conducted in Zulu. These portions occurred when guests of the show or callers to the show spoke to the hosts in Zulu. When the presenters and guests respond in Zulu to callers, it destabilizes, for a moment, the presumptive audience for the show; in tandem it destratifies the discrete, bounded language communities for Black South Africans. This feature meant that the program did not elide the tensions of language hierarchies within Black South Africa; many of my Sotho-Tswana speaking friends acknowledged that speaking a major Nguni language (either Zulu or Xhosa) was a requirement for negotiating Jozi, while Nguni speakers often did not have to acquire similar proficiency in Sotho-Tswana languages. What watching *Motswako* revealed is that the mixes it created, while apparently unintentional, reflected the experience of being urban, Black, and South African. The show has at different moments featured mixtures of class, race, and language that speak to the variegated present of South Africa.

The other space where motswako has come to prominence is in the music of post-apartheid South Africa, mainly in hip-hop and kwaito. Motswako has come to specify a genre of music associated with Mafikeng, the capital of the Northwest Province. More specifically, it is associated with hip-hop artists who mix languages (generally Setswana with English) and musical sounds (generally hip-hop, kwaito, house, and R&B). In fact, the two best-selling South African hip-hop artists of the 2000s, Pitch Black Afro and Skwatta Kamp (neither of whom are from the Northwest Province), while not referring to their music as motswako, have often been criticized for compromising hip-hop aesthetics with the inclusion of kwaito beats. Therefore, while it may have been Mafikeng hip-hoppers who coined the term, the notion of motswako has been in existence for some time.

Tuks, one of the more celebrated members of the hip-hop movement of South Africa and a proponent of motswako (as a musical style), states, "Tswana hip-hop is blowing up because it is proudly South African, as it is not scared to utilize kwaito or house elements" (Marumo 2006). Marumo writes that Tuks is not concerned about musical classification and takes on those who "condemn MC's who flow in their vernaculars or over unconventional hip-hop beats" (2006). Morafe, another of the hip-hop groups of that era from Mafikeng, claim, "Our sound [is] quite distinctive and cannot be boxed, because of the many influences our music adopted, like kwaito, funk, R&B, soul and hip-hop. It's a mixture of different sounds and languages, hence we call it '*motswako*'" (Seekane 2006). This mixing of musical styles, languages, and cultures is representative of South African public culture. If there is anything new, it is the acceleration of the mixtures due to media technologies and global circulations. Whether it be the language of S'camto (urban vernacular slang) or the music of motswako, there are local ways of explaining the outcomes of post-apartheid encounter that are useful alternatives to generalized terms of mixture. These local explanations ultimately speak to divergent processes. If we return to the club scene in Rosebank and reexamine it from the perspective of motswako, alternative readings emerge. The club was absolutely the site of a racially mixed scene. At the same time, it was also a place where Black working-class people could enter (with conditions, of course) to write themselves into the cultural landscape through embodied performances of complexity and contradiction. Similarly, the mixtures that enticed my township-based friends often had little to do with the presence of white clubgoers, and much more to do with the possibility of mingling with the Black middle classes and Black celebrity set.

"Freedom is being able to go where we want and be ourselves when we get there" (personal communication, 2004). This quote from one of my friends from Soweto highlights some key themes in the forthcoming encounters. Chiefly, I point to mobility and the collapse and reconfiguration of particular social relations through nightlife. In addition, I highlight how such circulations of mobility occur Afrodiasporically, as well as the ways in which music functions to satisfy Afrodiasporic longings. Emphasis is placed on the variegated way that the city is enjoyed by those pursuing sonic pleasures, the connection of those pleasures to social processes, and how encounter produces unexpected and differing outcomes. These encounters reveal the possible limitations to this flexibility and the unequal access to this mobility, as well as the fact that the freedoms of and within Afrodiasporic Space may be fraught with contradictions and tensions.

Encounter II: Space

Music has not only been central in the city's formation. It has contributed to its high levels of social energy which made music permeable, flexible, and defiant. — TANJA GERSHON

Encounter II takes place in the trendy bar and restaurant area of Melville. When I was searching for a place to live, I deliberately chose this area for its nightlife, bars, cafés, and the ability to walk the streets, which is unusual in Johannesburg. It is also an area close to major universities. The neighborhood has a reputation for being quirky and liberal, and most of its residents are white and relatively well off. In fact, the racial composition of the neighborhood has changed little in the immediate post-apartheid period in which my research occurred. As a result, while there are always Black faces in the clubbing and restaurant district, and the space is well integrated, it is not necessarily an interracial space in Melville. Hence, there are but only so many Black people who hang out and party in Melville on the weekends. Many of the bars and clubs have only a few Black patrons, and only a select few will have a more sizable Black crowd (approximately 33 percent of the crowd). There is also a large student crowd, which constitutes much of the Black patronage of the bars and clubs, depending on the time of the month. Like many people, students live from paycheck to paycheck in South Africa, often having their most disposable income at the end of the month.

I arrived in Melville with a group of friends from Soweto. They included two young men and two young women. Among them was one heterosexual couple; the other two were single and both were queer. I had met the group earlier in the day for lunch and we had decided to hang out for the day. Since they did not have their own vehicle, I would be doing the driving for the night, and most likely, I would drive them home at the end of the evening. The 7th Street drag (the main street in Melville), while busy during the day, is even busier at night. Whenever I bring people to Melville, I am always interested in what types of choices they will make about which bar or club they would like to enter. A few weeks prior, I had brought a group of friends from Durban to hang out in Melville; this group deliberately chose the one place that had a significant number of Black people among its clientele. I naively expected this group to do the same thing. As we were walking by one of the clubs, I saw that the entire clientele was Black, which is unusual for Melville standards. In addition, the party was being sponsored by Metro FM, the state-owned broadcasting channel whose target audience is urban-dwelling Black South Africans. Imagine my surprise when one of my companions turned to me and said, "This place is so ghetto. It's just like Soweto. Why should we leave Soweto to go to a place that's just like Soweto? No, thanks."

I was quite taken aback by the assessment of the club. Yes, the clientele was Black. The music was also the kind of music that one would hear in minibus taxis or at house parties around the townships. But I did not understand what was necessarily "ghetto" about it, unless, for him, ghetto was synonymous with Blackness. "We want something different," my companion explained. "We can go to places like this, hear this music, and see these same people anytime we want. What was the point of coming to Melville to hang out with a bunch of people that we could easily see in the township?" I disagreed: many of the people at the party appeared to be folks I had spotted in Melville before. I knew that some were my neighbors, and this was hardly a township crowd. Yet what I did notice was that while the crowd might not have literally been from the township, there was definitely a desire to re-create aspects of the township in Melville space. Several cultural critics (Hansen 2006; Nuttall 2003, 2004) have noted the ways in which young Black, upwardly mobile people have approached entrance into formerly all-white areas. These scholars suggest that the entrance into these areas is as much about shifting the contours of those areas, as it is about announcing a confident Black identity that is unapologetic about its township and working-class origins.

Eventually, after walking around with my companions, I decided to take a short drive to the gay club Oh! Oh! was located in another part of Melville. Oh! was never a particularly diverse place, with its clientele overwhelmingly white and predominantly male. However, it was definitely different from anything that one would find in Soweto, and it was also a distinct place for Melville. The upstairs is darkly lit, with a slower tempo of house music. The mezzanine floor features an outdoor deck for people watching the main bar, and a wall covered with televisions generally showed videos of white men in various stages of undress. The lower floor contained the main dance floor and was packed on this evening with sweaty bodies, grinding to the up-tempo, high-energy beat that is characteristic of techno-house music. After gaining entrance and looking around, they decided this was where they wanted to stay for the evening. Speaking to me, the young man in the heterosexual partnership said, "You know, I really love rave music." "I used to go to these rave parties a few years back; I haven't been in a while." His girlfriend remarked, "This club is nice; it's differently decorated." The gay young man interjected and said, "It's gay, so I'm cool with it."

For me, the experience at Oh! was so similar to the many predominantly white gay spaces in the United States. As an African American, I was annoyed at what I felt was the re-creation of a white gay space on the African continent that inhibited me from being exposed to the various ways that Black South Africans might experience queerness. Frankly, I longed for this space to be recast, to be made Black, and therefore, in my eyes, to be more representative of Africa. Yet what I was being confronted with was the reality that in the midst of all of these white gay men, we were indeed experiencing an aspect of African queer life, one that was no less real for those who lived it. Also, for Black subjects from the township, clubbing in Melville was not an everyday experience. It may very well have been a representative experience for me since, as a middle-class African American, I had access to these spaces regularly. This different type of experience was something that my friends wanted to have, but they did not have the kind of regular access that I had. The idea of my township companion participating in the overwhelmingly white rave-culture scene of South Africa had perplexed me and taught me something about unexpected routes between the space of identity and musical consumption. It also was the case that perhaps this crew of friends did not want to perform or be "township" on this evening. Unlike many at the places we went to (including myself), they would actually return to the township at the end of the evening. Their re-creation of new experiences for themselves, through radically different spaces and musical cultures, evidences the ways in which nightlife offers opportunities for self-fashioning. Each of us, after all, was involved in practices of the self: I wanted an explicitly Black space and was less concerned with politics of gender and sexuality, while two of my companions sought a more explicitly queer space, and they became less concerned about issues of race. In the case of the queer Black woman in the group, the reconfiguration of gender offered a liberatory site of performance. The heterosexual couple enjoyed the space because the woman loved the decor and "classiness" of the space, and the man appreciated rave music.

This encounter speaks to the fact that there is no such thing as a value-neutral space. Space contains meanings that are intimately connected to notions of race, class, gender, and sexuality. These meanings determine who enters, how they are treated once they enter, and whether they return. Initially, this was my concern about our group going to Oh! I felt we would stick out, perhaps not even be allowed in by the doorman because we were a group of young Black people who were not marked as queer, at least not in ways that would be visible to the crowd at Oh! Thankfully, on this occasion, this was not our experience. Yet this does not mean that Oh! was never accused of enforcing a racially discriminating door policy, or that such policies are not tacitly practiced in some of Jozi's (queer) club spaces. In addition, while Oh! does not require a cover charge to enter, its drinks are more expensive than most clubs. These facts speak to the possible limits of this particular space's ability to provide the forms of self-fashioning that I posit. Nevertheless, in this encounter, my friends had a great time because they were experiencing something different, something that simultaneously marked them as from the township but not completely defined by it. My companions deliberately avoided spaces that reminded them of "home" in order to redefine themselves within the township of their everyday experience, as well as outside of it. I became aware of this only later, when I was interacting with a separate group of friends and was told repeatedly about how much my companions for that evening had spoken about their night out in Melville.

As Katherine McKittrick argues, Blacks in Afrodiasporic Space experience geography as an interplay between geographies of domination and their own experience of space (2006, x). For McKittrick, "geography is not, however, secure and unwavering: we produce space, we produce meanings and we work very hard to make geography what it is" (xi). As the bodies of mostly young Black people move throughout the city seeking sonic pleasure, there are implications of their presence in particular spaces. Such spaces are open, closed, or, in the case of Oh!, ajar. The city is the ultimate expression of possibility for the human. It is the space of the oeuvre or the opening through which the innumerable moments of the everyday combine to create possibilities. The city is after all made up of several dimensions. These dimensions not only create space for the state to impose its vision onto the urban, but also create both stark oppositions and possible connections. We are reminded that "while the power of transparent space works to hierarchically position individuals, communities, regimes and nations, it is also contestable—the [Black] subject interprets and ruptures the knowability of our surroundings," thus remaking the spaces of the city (6). Thus, I am most interested in the city as a place of encounters. Within these encounters, the city becomes a place of desire, permanent disequilibrium, play, and the unpredictable (Lefebvre 1996, 129). The reframing of space through nightlife and the practices of partying speaks to experiences of marginalization but also to the labors of self and community that allow for the appropriation and reappropriation of city spaces (Livermon 2014). These processes are intimately connected with how the "racialized, gendered, sexed, classed and imaginative body-self necessarily interprets space and place—in its limitations and possibilities" (McKittrick 2006, 2).

These spatial encounters also unfold in the context of increased mobility. While my friends cannot live in these expensive areas of the city, they freely party in them. Mobility of the Black body is precisely what apartheid attempted to curb: it was predicated on the literal and figurative static Black body. Any mobility of the Black body was connected to the ways in which that body could service the needs of racial capital. In the post-apartheid period, the types of mobility are opened up, not only by the rise of the minibus taxi, identified by Hansen (2006), but also by the rise in access to private vehicles. It is also the case that such mobility of bodies is not a one-way process: musical mobility works to deliver township into the space of the posh suburbs, and it also delivers the suburbs into the space of the township.

Encounter III: Mobility

I saw a township that was alive. On the road after 6pm was evidence—traffic cars were obviously going to and from somewhere. I don't know where but the amount of traffic on the roads on a Saturday evening showed that people were moving. . . . Also there were a lot of local taxis on the road—"amaphela" as they are popularly known. . . . [T]his meant the community was active in social transactions albeit at night and on a Saturday.—WANDILE NGCAWENI

The practices of self-fashioning, urban experience, and emerging identities in post-apartheid Johannesburg would be impossible without the new sense of mobility that was unfolding in this era. This mobility unfolded in the pursuit of sonic pleasure as township tastes, styles, and bodies seeped into formerly white areas. These improvisatory ways of being in urban space also corresponded to the infusion of the townships with urban dwellers who sought not only a novel musical experience, but also a social one. I introduce Encounter III to delineate the characteristic of middle-class mobility: not only did working-class township subjects bring the mix, but so did middle-class, suburban subjects. The key substance of these encounters is that of longing: Black South Africans in transit—geographic, economic, social, sexual—sublimated the growing tensions between what they may perceive as their "roots" and their current suburban lifestyles in the textures of the club. For them, the township becomes an Afrodiasporic Space of resolution, if only temporary: a way of navigating the issue between the upwardly mobile subjectivity and identity that would forever remain tied to township life. Afrodiasporic is used to describe these relationships because the longing for the township as a "home" space produces feelings of nostalgia for ex-township residents. In this sense, Afrodiasporic theory reveals the relationship between the melancholia of longing experienced by the newly middle-class South African Black subject, and members of the global Black diaspora, whose sense of displacement, self-making, and contradictory modes of Otherness amplify the post-apartheid movement, even as the genealogies that produce these displacements are distinct.

For my friend Mpho and me, a night out in Soweto was always a special treat. Mpho was the first in his family to complete a tertiary education: his family was solidly middle class, and both of his parents were teachers. Raised in townships throughout Gauteng, Mpho was sent to Swaziland to complete his high school education during the turbulent years of the 1980s, when numerous protests and boycotts made attending township high schools nearly impossible. In the early 1990s, Mpho was among the first generation of young Black South Africans to move into central Johannesburg. He was also among the first large wave of Black students to attend the University of Witwatersrand in the early 1990s. Mpho recalls with fondness the days of dancing to house music in central city areas such as Hillbrow and Yeoville, as well as his days of living in a cramped apartment in Berea. He notes that now most of his friends, like himself, enjoy spacious homes in the northern suburbs of Johannesburg or the newly gentrifying CBD neighborhoods; they drive nice cars; and they make good money. "We are a long way away from those days of living in cramped flats in the CBD," Mpho said with a laugh. I met Mpho years ago, and we formed an immediate friendship. While he is well educated and bourgeois, Mpho still kept many of his township connections; in turn, he helped me navigate the nexus between the city and the township.

On one particular evening, we were joined by two friends of mine from Durban. One was African American, like myself, and the other friend grew up and lived in Umlazi, the major township in the Durban metropolitan area. That evening, we had a choice of attending either a street bash or two different clubs, the Rock and Meli's. Since it was early in the evening, we decided to try to find a club that we had heard about that was located in Protea, one of Soweto's nicer suburbs. I was irritated because I was not only unable to find the club, I was also lost. Then, I was disturbed further because I saw the police flashing their lights in my rear mirror.

The young people of Johannesburg will tell you that police often stop motorists (and young Black males in particular) unprovoked, with the hope of getting a bribe. These stops do not legally require signs of illegal activity; in fact, young Black male drivers must simply factor in the forces of police corruption and surveillance into their everyday lives. When you are a student researcher with US citizenship, the corruption takes on a threatening dimension: a routine traffic stop can quickly turn into a night in jail, and it also carries the threat of deportation. Black bodies bear the brunt of this kind of corruption primarily because regional practices of policing-by-surveillance were a dimension of colonial-apartheid social control, meant to constrain the movement of bodies that looked like mine. As Simone Browne reminds us, surveillance practices that developed from slavery (and further, in the South African case, colonialism and apartheid) were a "violent regulation of Black mobilities" (2015, 52). "Where are you from?" an officer asked me in a gruff voice, even after I handed him my passport and driver's license. I tried to temper my irritation at this question, so easily answered by my paperwork. It was hard for me to tell if this was just typical police rudeness, or a failed diasporic connection: a rupture between my own Blackness and the specificities of nationality and ethnicity that threatened to shatter it with another round of misrecognition. I find that, in these cases, speaking to the officers in Zulu tends to hasten their irritation with me as it displaces and disorients the body they see in front of them. In the breaks, in these spaces of in-between for these officials, my Zulu was too good to belong to an American, but too bad to belong to an African (Moten 2003). My English, they figured, was not stereotypical enough. These coded assumptions meant that I did not possess the stereotypical African American vernacular accent associated with Black Americanness. After looking at me closely, my passport picture, and the picture on my driver's license, the officer finally let me go. I could see that my presence was discomforting. I was not who or what the officer expected me to be. As a queer African American who was working toward a PhD, I was used to my presence causing discomfort, even when I was at "home" in the United States. I am used to living in the slippage between preconceived categories of nation, masculinity, or race.

Thus released, I was happy to be free of the police and the anxiety that any encounter with them causes; we continued on to our destination. However, it is worth mentioning that because of the dominant representation of the young Black men as inherently dangerous and pathological, the relationship between Black men and the police is riddled with anxiety, something that class and national status cannot always overcome. It is an encounter: it is my encounter as it resonates with the precarity of in-betweenness that my South African experience creates. In this register, it makes the work of liberatory self-writing in the club that much more crucial.

When we finally arrived at the club, I realized that, in a way, both Mpho and I participated in Afrodiasporic Space, returning to the "homespace" within its walls. Yearning, after our police encounter, to move in and through a crowd of Black bodies, we had abandoned our search for the other club and settled on going to the Rock. Mpho explains it this way:

> So, during the work week, I am surrounded by a mostly white environment, right? Then I go home to my mostly white neighborhood. When the weekends come, I just want to be in a Black space. For so many reasons I can't actually live there anymore, I mean the commute to work, the space, and the privacy. But that doesn't mean I don't need the experience of the community. In fact, I need it more now. What better Black space can there be but Soweto? When I want to have my township with a touch of class, then I head to the Rock.

I share Mpho's sense of yearning, even if my Black Americanness puts me in a very different relationship to the club. While my removal from the space of Blackness as it pertains to South Africa is more extreme than his (unlike in West Africa, I cannot even claim ancestral roots), I still have a diasporic experience in relationship to a place like Soweto. This is so if one keeps in mind that Diaspora is about the experience of and affinity with Blackness, not a set of genealogical or genetic markers.

As a child, I wrote an essay that I found some years ago while cleaning out an old desk. I wrote it as a seventh grader in the mid-1980s during the time of increased uprisings and activism within South Africa, and during a time of increased US media attention on South Africa. I wrote of how I looked forward to the day when South Africa would be free, and how it pained me to know that there were still places where Blackness automatically meant second-class citizenship. Like many African Americans who grew up in the 1980s, I identified with the struggle against apartheid, perceiving the injustices of the system to be related to my own experiences of provisional citizenship in the United States. For me, Afrodiasporic Space was produced in the moment of desiring to be in Soweto, that place that represented the struggle, and to know that I was free to move into the space of Soweto and engage as much as possible with the people who live there. While this was not a journey to the slave castles on the West African coast, it was no less diasporic, for the Afrodiasporic Space created by the fight against apartheid mitigated my understanding of the importance of Soweto and the meaning behind my social interactions with its inhabitants.

At one point during fieldwork, I stayed with my boyfriend in Soweto for about a month. I explained the experience to Mpho: "I have to wake up so early just to get to school by 9! First, there is the rushing to the outdoor toilet in the freezing cold of the morning. Then there is the quick dash back to the house where I begin boiling kettles of water so I can have a bath. Then there is the waiting for the taxi to fill up before it can proceed to town." Mpho responded by saying, "Well, you are getting the real township experience. We all remember it, and we don't want to go back to it." I realized at that juncture that while I could live in Soweto, it would have to be one of the areas of the township that more resembled the suburban areas of Johannesburg. It was uncomfortable to admit that, as an African American, I could not really live in this space the way hundreds of thousands did on a daily basis without altering it to suit my needs. Nevertheless, I felt somewhat ambiguous because the township space is always in flux, and throughout townships such as Soweto, amenities are being upgraded on a daily basis. My desire to occupy one of these spaces, and perhaps Mpho's desire to as well, should not be read necessarily as wanting to refashion it, for there was nothing inherently *township* about an outside toilet, just as there was nothing inherently *suburb* about hot running water. In contrast to Mpho's statement, my experience was not *the* Soweto experience; it was simply *a* Soweto experience. My desire for Soweto, I realized in that moment, was no less or no more romantic than Mpho's, yet our experiences of displacement from the township came from radically different trajectories. Each of our subjectivities was produced from a unique tangle of identification, desire, and displacement—not only physical, but also psychic and political complexities.

Located in Rockville, near Thokoza Park, and not far off of Chris Hani Road (the main street of Soweto) the Rock is well situated in terms of Soweto topography and is relatively easy to find. Apparently, the club started off rather inauspiciously, as a regular home converted into a place for partying. As clientele increased, owners expanded and upgraded the space. The club now boasts an upstairs roof area, an outdoor patio area, a separate smoking section, a DJ booth, a dress code, and a back room. While the Rock is more upscale than similar places in Soweto, little effort has been made to dress it in fancy decor on par with upscale Johannesburg and newer Soweto clubs. There is no cover charge, but the club's drinks, while relatively well priced compared to venues elsewhere in the city, are still pricey by Soweto standards. For this reason, the club has earned a reputation for hosting the Soweto bourgeoisie and the aspirant class. Despite its upmarket reputation, the wide range of cars in the packed parking lot betrays the mixed financial backgrounds of its clubgoers. That evening, the cars ranged from rather inauspicious Citi Golfs and Beetles to late-model BMWs and Mercedes.

The Rock is an interesting place because of its ability to draw one of the more diverse crowds in the Jozi scene. In fact, the presence of white, coloured, or Indian South Africans rarely elicits much of a comment or a second glance. Celebrities from local soap operas, soccer, and kwaito mix with regular folk. It is not unusual to be dancing next to the actor who plays in *Generations* and his entourage on one side, while a group of gays and their friends gyrate in a circle behind you, and a group of *majita* are on the other side. On one particular night, my visitor from Umlazi was surprised to see the presence of Coloured trans women/cross-dressing men in the space, given perhaps the different possibilities for visible gender and sexual transgression in Durban townships versus Johannesburg. The visitor from Umlazi said, "You come to Jozi and you see all kinds of crazy things. I would never see anything like this in my township. A man dressed as a woman!" Mpho and I turned to him and said, "Welcome to Jozi." Interestingly, for my Durban visitor, the issue was not the person's racial background, but rather that she was trans/gender-nonconforming, and his reaction was more curious than dismissive. What is perhaps equally noteworthy, however, is that trans women/gender-nonconforming men caused no commentary or second glances, except from my friend from Durban. They were just a part of the nightlife scene, dancing with friends, grabbing drinks, enjoying the space. In fact, the Rock during that time earned a reputation in the queer community of Soweto, and consequently in the larger queer community in Jozi, as a queer-friendly space. It was not unusual, then, to find this space that is actively produced as Black and heterosexual facilitating other kinds of interactions. These interactions can be between foreign tourists and local Soweto residents, between straight and queer Jozi residents, between the wealthy and the working class, or between Black and white. I do not mean to suggest that the Jozi party scene or the Rock is a place of uncritical celebratory multiculturalism. While the space exists in one particular realm, it also provides possibilities for other types of encounters, other forms of mixing. In this sense, the Rock was important to the Soweto and Jozi scene because it allows for spontaneous and radically diverse interactions.

Gilroy (2004) calls these diverse interactions moments of "multiculture," and they are significant because they counteract the desire for melancholia, particularly the melancholia that is built from the maintenance of white supremacy and racialized exclusion. The mixtures in Jozi's club scene also fit within Gilroy's concept of "conviviality," as they share the dimensions of cultural conversation Gilroy calls the convivial nature of postcolonial cities. While more explicitly racialized moments do exist in Johannesburg, the encounters of conviviality that concern my study arise from interactions between white and Black in Norwood, or the increasing presence of white, coloured, and Indian South Africans in Soweto nightlife. Spaces of conviviality are ephemeral, but they must not be taken for granted. Their existence, while complicated, offers a way to understand why such cultural practices and spaces retain importance for post-apartheid South Africans.

After dancing for a few hours and having a couple of drinks, Mpho, our Durban visitor, and I realized that the crowd was beginning to dissipate. I glanced at my watch, and noticing that it was nearly two o'clock in the morning, realized that I had never actually stayed at the Rock this late. In fact, the bar was getting ready to close, a phenomenon that I had never experienced. At the entrance, people were barred from entering the closing scene. We left, determined to continue our partying at Meli's. When we arrived in Pimville, we were surprised to see the number of cars outside of Meli's. The calm, seated atmosphere we witnessed when we had driven by earlier (on our way to the Rock) had transformed into something else entirely: everyone who was at the Rock, it seemed, had left and gone to Meli's. If one could say that there was such a thing as an after-hours place in Soweto, this would be it. While the dress code at Meli's was unclear, the entrance policy was similar to that of the Rock, with body searches at the entrance. From the outside, you could hear the bass of the music thumping, and there was the usual assortment of folks (car-park guards, hangers-on, people wanting to finish their drinks) outside. Once inside the club, it had been transformed from the relatively sedate place it had been earlier in the evening. The DJ, who had been playing far more relaxing jazz and soul-inflected sounds, had switched to house and kwaito. I imagined that some sort of fire code must have been violated, as I could not understand how so many people could fit into such a small space. Many, but not all, of the tables had been pushed to the sides to enlarge the dancing area, which earlier in the evening had been nonexistent. And before that moment, I had never seen a club so devoted to the art of dancing. I witnessed what was probably the most intense kinesthetic performance to Lebo Mathosa's "I Love Music," aside from that of the artist. The remix, played by the DJ, emphasized the backing house beat but also elongated the piano riffs of the song and punctuated the high-hat intervals. Mathosa's voice, loungy and languid, yet guttural and fierce, took on an otherworldly quality.

The combination of the house beat (perhaps a bit faster than most kwaito) merged with the melodies of voice and the percussive piano elements to send the crowd into a frenzied state. It reminded me that South African dance floors often saved their most intense celebrations for beloved local artists. All pretense and posing was left at the door as young men, women, and even a few older folks got down to the music. This space was dedicated to the principle of uninhibited dancing, and whether I wanted to or not, I was caught on the crowded dance floor jostling for space, singing the lyrics, timing my moves as much with the orgiastic quality of the crowd, which in unison seemed to understand the beat and move collectively through the song.

> Beauty in your eyes
> I see love
> The way you move your body
> I see music
> For better or worse
> I will love
> Love music
> In sickness, in pain
> I will love
> I love music
> Just tell me what's your secret
> 'Cause I see love
> The way you move your body
> I see music (Lebo Mathosa, "I Love Music")

In this moment, through the song, and through the space, the young people gathered together and produced a collective joy, an eroticism of self-celebration and collectivity. The lyrics in many ways help to create the experience of love, what it means to see beauty in the eyes of your dance partner, to see love as they move their bodies, to see their bodies and their beauty as music. Even though I know I was there, years later, glancing at detailed fieldnotes, I am still unable to grasp fully and describe the multiple senses of that spectacular moment and realize that my attempt to reproduce it in written form can occur only with significant conceptual limitations. And yet I am drawn to the moments after the spectacular heightened moment, when you realize that you just participated in and witnessed something special, the tingly joy of adrenalin being released from the body as heightened sensation dulls. In subsequent years, I have thought more about the inversion of senses that Mathosa proposes in the song. How might we "see" music, and how does seeing music allow us to have an experience of joy, of love, of eroticisms of collective behavior? The kinesthetic movement of collectivity that I describe happening at Meli's as we translated our individual experiences of her song into movement became a form of seeing: feelings become material, attitudes become sensory, self-conception becomes real. Our individual translations of sound into visible kinesthetic movement created a collective response as we attuned ourselves to our partner(s), to the people surrounding us, to the DJ. In that moment motswako—an encounter that mixed senses, sound, perception, and bodies—occurred.

The City Remastered

Johannesburg is one example of how Afrodiasporic Space is produced through encounters related to post-apartheid nightlife. Dancing, mobile, impactful kwaito bodies use their performance practices to lend meaning to space in the city, in essence creating a "performance geography" that remakes the city even as it engages Black musical forms throughout Afrodiasporic Space (Niaah 2008). The city, for many in Afrodiasporic Space, becomes a place of hypercontestation, particularly around the bodies of younger Black people. Leisure practices formed through and across Afrodiasporic Space help create the *metswako* (mixtures) that mark post-apartheid South African society. The social encounters illuminated here work diasporically because young Black bodies respond to their exclusion from city space by drawing on the music and the affects of the African Diaspora. Writing about city life, AbdouMaliq Simone (2009) argues that similar forms of quotidian creativity enacted by people of African descent worldwide constitute a kind of "Black urbanism." Critical for the understanding of the intersections of space and sound, Simone (2009) argues that "Blackness does not constitute a particular kind of urbanism, but rather tries to bring into consideration certain dimensions of urban life that are too often not given their due" (278). In relationship to music in nightlife, this dimension of urban life is firmly situated in forms of cultural creativity that work within and through the realities of contemporary neoliberal globalization. Similar to the inner-city youth of color chronicled by Robin D. G. Kelley (1997), Black South African youths' pursuit of leisure is in no way divorced from labor. Nightlife, examined here, reimagines the bodies of young Black South Africans. It is a way of fighting back against the confines of labor: the transformation of the spaces of lack and isolation become through sound reimagined as nodes of interconnection, the periphery becomes the center, spaces of stasis become sites of mobility. In addition, such practices, by challenging norms and conventions of race, gender, sexuality, and class that circulate in relation to Black bodies internationally, work to create the space for Black bodies to be imagined differently. The joys that erupt through experiences like this evidence the erotic possibilities of nightlife.

The political dimension of club life becomes clear through the concept of the constitutive outside: the spatial dimension of Otherness as it relates to post-apartheid Johannesburg's pockets of encounter. Historically (and one might argue, contemporaneously, through the discourse of illegal African immigration), the body of Blackness has been configured as the destabilizing, dangerous external presence through which the (white) South African state was established. The policy of grand apartheid sought to render the Black population without citizenship by enacting a homeland policy, meant to keep white spaces absent of Black people. This spatial designation was reiterated in the urban areas with the creation and enforcement of pass laws. Therefore, the Black body not only was the constitutive outside of the state, but, through the homeland and pass policies, was also imagined as outside the city. This was rendered spatially through the creation of townships, and ideologically, through a discourse of premodern tribal Africanness marked by rural, unchanging ethnolinguistic traditions. During the apartheid era, the horn at Turbine Hall reminded Blacks that they had to be outside city borders unless they had special permission to be present. In an interesting moment of rewriting space in the post-apartheid era by the mid-2000s, Turbine Hall became one of the sites of new musical expression hosting hip-hop and kwaito parties and being the location of the hip-hop group Skwatta Kamp's "Clap Song" video (2004). By 2009, the "redevelopment" of Turbine Hall made it a far cry from the edgy underground music space that hosted a variety of established and up-and-coming acts in the early 2000s. As one of the city's renovation projects, Turbine Hall hosts upscale events, predominantly weddings and conferences (see figures 2.2 and 2.3).

The task of the post-apartheid government has been to restore the Black body both to the idea of the nation and to the place of the city. The scrapping of

FIGURE 2.2. Historic Turbine Hall, c. 1923, prior to modern-day renovations. Source: www.theforum.co.za.

FIGURE 2.3. Turbine Hall today. Source: Atterbury Holdings.

apartheid-era laws has thus given Black people the right to the city. The city of Jozi has become a place of possibility for those Black young bodies that were so perpetually excluded. What emerges is an understanding that the right to the city is not simply a one-way process. In this way, the geographies of the city are remastered, not completely remixed or reformulated, but enhanced through the original logics. The examination of space, sound, and the body in this chapter reveals that the changes that have occurred in the post-apartheid landscape of the city are not unidirectional or centered. Rather, the pursuit of sonic pleasure shows a city that increasingly assertive, mobile young Black people traverse from township spaces to predominantly white suburbs, accompanied by whites, Coloureds, and Indians who no longer regard the Black townships as no-go zones. While new musical cultures highlight the push of township into town, they also highlight the push of town into township. Kwaito is not simply of the townships or the suburbs, it is both and neither. Attempts to situate kwaito in one area or to contain it in one place says more about the attempts of various actors (the government, multinational corporations, parents) to place it outside: to control and surveil young Black bodies. They are incommensurate with the encounters through which young people experience and imagine the intermixtures of urban space. Those people who have moved out do not simply sever ties: they choose to revisit the sonic arenas of the township for both the music that is played and the atmosphere that it provides. Likewise, those people in the township seeking out new and different experiences enter into the havens of the party scene of the (Black) bourgeois. This creates a (dis)connect between the increasing mobility of young Black people in the city of Jozi, and the ways in which postcolonial understandings of interior and exterior spaces attempt to discipline and organize that very mobility. Amid these tensions, the city that is made anew is one where the connections between the city itself—the "town" and its outlying areas—and "the township" thicken and multiply. The tropes that consistently mark these areas as completely separate in experience—town as white and wealthy, township as Black and poor—are no longer valid.

As a result, kwaito performance works as a mode of organic intellectualism that springs forth from township spaces to express various types of emotions, desires, and feelings from a class of people that often do not have the power to express and represent themselves (see Peterson 2003). These people enter the city confidently on their own terms. Yet the experience of kwaito also indicates a growing Black middle class, and the concurrent incorporation of Blackness into the city. Bourgeois spaces of integration and equality allow for a new form of Blackness to flourish: one that emplaces itself in the form of middle-class residence (see Nuttall 2003, 2004). Kwaito can be about the relationships between the emerging Black middle class and the township, as well as spaces "elsewhere" through the engagement with Afrodiasporic sounds. Mixtures, sometimes unexpected and at other times ephemeral, define this space. Overall, the post-apartheid city in formation displays the residues of apartheid, simultaneously creating a space for the radical reimagination and obliteration of the apartheid logic.

3

"Si-Ghetto Fabulous"

SELF-FASHIONING, CONSUMPTION,
AND PLEASURE IN KWAITO

How do the performances associated with kwaito engage or disengage the social hierarchies that constitute South African society? How are the intersecting positionalities of race, class, gender, and sexuality reproduced, repudiated, reconfigured, or reimagined through sonic, poetic, and corporeal performance practice? These questions form the interrogatory foundation of this chapter in order to not only situate kwaito within the social economy of South Africa but also illuminate the inherited models of freedom that kwaito aims to produce and circulate. My aim is to show how the sonic geographies of kwaito within the Johannesburg landscape reflect, amplify, and reconfigure both the everyday life of the city and its relationship to the African Diaspora. In the rhythms of the music, the everyday patterns of Johannesburg living are spun into different rotations: different relations and configurations. While race, class, gender, and sexuality come to bear on the politics of kwaito space, they are also reconfigured within that space as kwaito practitioners imagine Johannesburg anew through the practices of partying, nightlife, and sonic pleasure.

Boundaries are sustained and traversed; on the dance floor, hierarchies are elided. In order to advance this analysis, it is important to foreground the politics and dynamics of consumption as it relates to kwaito practices. Conventional journalistic criticism of the scene takes for granted the notion that kwaito and its attendant cultural practices manifest a youth culture obsessed with materialism and divorced from civic engagement. However, in the breaks between scene and product, city and subject, is a world of possibility: a field of self-making and self-writing by which kwaito practitioners harness a political presence not usually or readily available to them in the public domains of social life external to kwaito performance realms. In looking at the opening of Maponya Mall in relationship to the preparation for a kwaito festival, as well as the controversy created by e.tv's exposé of s'khothane culture, kwaito bodies perform their relationship to the music, and to the politics the music animates. Party preparation as a site of kwaito performance is the focus. I explore the complex of social exchanges—the patterns and performances—that make the mobile evenings previously described possible. Pleasure and consumption are vital conceptual categories that illuminate the politics of the kwaito body: in lieu of the disease, violence, and poverty associated with township life, I locate remasterings of freedom. In the context of careful, participatory ethnographic engagement, the moments I analyze reveal consumption to be a critical practice of self-fashioning and citizenship, and an important analytical tool. Ultimately, kwaito cannot be reduced to a politics of surrender because as a politics of exuberance there is always an excess that escapes capture.

Encounter IV: The Opening of Maponya Mall

In 2007, Nelson Mandela; the mayor of Johannesburg, Amos Masondo; Sowetan businessman Richard Maponya; and a host of South African celebrities braved the damp and rainy early spring weather of Johannesburg to join approximately 100,000 shoppers (including myself), to witness and participate in the opening of the Maponya Mall in Pimville, Soweto. The mall represented the fulfillment of a dream for Maponya: despite numerous threats and legal challenges from the apartheid state, Maponya had managed to keep in his possession a large tract of land in Pimville. In the 1980s he wanted to develop a mall with the conviction that the millions being spent each year on transportation to upmarket shopping venues in Johannesburg could be better spent in Soweto itself. The apartheid government thwarted his chances to build the mall; later, the market and rand crash of the mid-1990s made it prohibitively expensive. Finally, in 2005, circumstances permitted Maponya, in partnership with a property development association, to break ground on the mall.

Practically, the mall means access to full-range grocery stores, banks, and other essential services that many had to travel outside the community to previously obtain. According to Mayor Masondo, "the coming into being of this retail facility in Soweto will result in household budget savings, savings in travel time and generation of local jobs in the retail sector" (Da Silva 2007), among other things. However, Mandela, a host of celebrities, and tens of thousands of everyday Soweto citizens were not present simply because the opening of the mall would essentially keep money in the community through the saving of both travel time and money and creating local jobs. They were there to witness the excitement in the air: the sense of possibility that superseded the possible financial benefits of the mall's presence in Soweto. Finally, Soweto would get its own upmarket mall, comparable to the malls of the northern suburbs that many Sowetans typically shopped at on a weekly basis, despite the difficulty and costs of getting there. This unequal access—an immense gap in consumer agency—was being rectified with the building of the Maponya Mall.

The conveniences of living in "white" Johannesburg were being brought to Black Sowetans, and the move represented a transformation of their role in the economy: a mode of subjectivity they had been denied under apartheid. Maponya himself summed it up best: "Today I brought Sandton City [universally recognized as the most upscale mall in South Africa] to Soweto and the onus is now on you people to ensure that you take good care of this establishment. This property is yours and your children's place, protect it" (Da Silva 2007) (see figures 3.1 and 3.2). The excitement that swirled around the opening of the mall was about more than the capitulation of Sowetans to consumer culture. It was also about the recognition that their rands were worth something, and the dignity of having facilities comparable to those routinely enjoyed by middle-class white South Africans in their own neighborhoods.

While conducting research at that time, I lived with friends in Jabavu, near the recently completed Jabulani Mall, which struck me as offering a wide array of shopping possibilities previously unavailable in Soweto. And yet with the opening of Maponya Mall, there was a great deal of excitement in the household. Most of the household members, particularly the older women, were excited about the opening-day deals that were being advertised by the major grocery retailers. They perceived the opening of the mall as an opportunity to restock their pantries at a reduced price. My friend Thato and I wanted to just be a part of the excitement, so although we did not approach the event with the idea of shopping, we were going to people watch, be a part of the

FIGURE 3.1. Opening court of Maponya Mall. Source: Redefine Properties Limited.

FIGURE 3.2. The shopping venues of Maponya Mall. Source: Redefine Properties Limited.

"event," and, perhaps if we saw a good deal, do a bit of shopping. Traffic to reach the mall was backed up all along Chris Hani Road, off of which the mall was located. As we inched forward to the guarded and secured parking lot, we realized that we needed to have left home far earlier than we did in order to reach the mall. By the time we arrived, the main dignitaries and politicians had already spoken and departed, so we were left to mill around the mall and see the stores for ourselves. I was immediately impressed with its architecture, which in its (post)modernity mimicked the upscale malls of wealthier parts of Jozi.

While much was made of the celebrities and the Black elites' appearance at the mall, I was struck by the everyday people who wandered about the open and airy spaces. By far, the largest crowds were not at the specialty boutiques (many of which had not completed construction in time for the opening) but at the grocery stores and in the open public spaces near the food court. This made sense, given that it was the end of the month when most people would have needed to do their shopping, combined with the aforementioned specials that were on offer. While we wanted to see if we could benefit from these sales, the lines were too long, the crowds too heavy to venture inside. Instead, we moved through the less-busy stores, checked out some of the restaurants, and visited the bank and cell-phone stores to withdraw money and buy airtime. My initial impression was that the entire event was rather mundane, perhaps not worth the effort to partake. However, as I began to grasp the complex historical and geographic circumstances that unfolded at the mall that day, it has become clear that retail space signifies something much more profound for the residents of Soweto, particularly its younger set.

There are numerous legitimate critiques of Maponya Mall, its opening, and what it has come to signify for post-apartheid South African urban planning. Certainly, the money used to construct and build the mall (even if privately financed) could have been used to address Soweto's more pressing needs: additional housing, stronger transport infrastructure, or more local health clinics. In the face of significant housing shortages (including a shantytown only a few kilometers away from the mall in Kliptown), the focus on additional retail space as a feature of development seems misguided. In constructing securitized, private space, city planners de-prioritized investment in more accessible noncommercial public space and capitulated to a neoliberal logic that equates consumption and citizenship. Subsequent studies have shown that Sowetan residents primarily use the mall to shop for necessities and that luxury and big-ticket items are still purchased primarily outside the townships, lessening perhaps the claim that this mall would equalize consumption for township residents. Furthermore, while Soweto residents have enjoyed some savings on transportation costs and time as a result of the mall's location, evidence seems to suggest that these benefits may have been exaggerated. There is also a question of who owns the stores and shops in the mall: to what extent have major chain stores—the kind exclusively owned and operated by white South Africans—displaced small local businesses? Preliminary studies suggest that Black-owned small businesses have suffered as a result of the proliferation of corporate stores in township space (Bushman 2012).

With all of these criticisms in place, however, the mall remains enormously popular among Sowetan residents in surveys. Why might that be the case? The history of colonial and apartheid regulation of Black consumption frames consumption as a liberatory practice. Rather than see consumption as a lamentable by-product that is contrasted with processes of liberation, consumption, particularly the notion of equal consumption, is attached to questions of freedom for many Black South Africans. Writing about the history of consumptive practices for Black South Africans, Deborah Posel reveals, "The making of the racial order was, in part, a way of regulating people's aspirations, interests and powers as consumers. The desire and power to consume was racialized, at the same time that it was fundamental in the very making of race. This interconnection, in turn, has had a profound bearing on the genealogy of varied and contested imaginings of freedom" (2010, 160). A central aspect of the colonial-apartheid order was that consumption was used to fashion racial hierarchies, and one's place in the system was intricately linked to one's ability to consume equally. Hence, Posel concludes, "The struggle against apartheid was in part a struggle not just to transcend poverty, but to become rich; that freedom was expressed, in part in acquisition" (158).

In her study of leisure practices as a specific form of consumption at the close of apartheid, Mehita Iqani argues that "the right to leisure [was] guarded by the white minority" (2015, 134). Analyzing a 1991 documentary that focused on a group of Black golf players at the end of apartheid, she interrogates scholarly assumptions that hold that consumption practices cannibalize liberatory discourses by sacrificing them for personal enrichment and conspicuous displays of wealth (140). While she agrees that many consumer practices can be ultimately self-defeating for Black South Africans, Iqani cautions that such a reductive approach to analyzing consumption in South Africa misses the practices of self-making that animate so much youth engagement with the market and its products. She writes, "Consumption is a set of practices in which agency can be exercised, confidence built, and the material pleasures of life accessed and enjoyed" (140). In what follows, the micro-practices of self-involvement in experiencing the Jozi party scene are analyzed. Here, Posel's (2010) and Iqani's (2015) concerns about the relationship between freedom and consumption are instructive. I link their concerns with an explicit call to recognize the pleasures and joys that inform the (erotic) politics of exuberance and excess. Following the work of Black queer scholar E. Patrick Johnson, excess is considered productively as that which "cannot be contained in conventional categories of being" (2001, 2). It is its unruliness that makes it politically productive and a site of anxiety for post-apartheid society. Consumption as a medium for the experience of pleasure and dignity contours my ethnography of a kwaito festival in the park. This approach reveals how carefully cultivated practices of preparation, space-making, and self-fashioning emerge from modes of consumption that are animated by the politics of pleasure.

Encounter V: Kwaito Festivals—Consumption, Self-Fashioning, and Pleasure

Self-fashioning is essential to understanding the realities of post-apartheid South Africa: it is one of the few spaces of control that people who are otherwise disenfranchised have over their own bodies (Mbembe 2002). For instance, the space of a park, reworked by musical leisure, serves as a site for exploring these moments of self-fashioning. The heavy policing of the park speaks to state anxiety brought on by the presence of young Black bodies that are also mobile in an open space. Festivals tend to be held in open-air places, in parks either in the township itself or inside the city limits. Depending on the sponsoring body, these festivals range from being free (when sponsored by the government or nongovernmental organizations) to charging a nominal fee (when sponsored by corporations) to being more exorbitant in price (when sponsored by a party promoter). In each case, partygoers are allowed to bring coolers with food, snacks, and drinks; therefore, these festivals have a tendency to have a picniclike atmosphere.

 I came to know the individuals who were planning on attending this party just a few weeks before it took place. While at Oh! in Melville, I met my friend Tumelo. I discussed my research with him and the fact that I was having difficulty meeting up with people who lived in nearby townships. As a diasporic African and as part of fieldwork, I desired to have close connections with township-based South Africans, but these connections were often difficult to come by spontaneously due to differences in class, language, and nationality. Tumelo was dating a bourgeois Black man, and he lived in the northern suburbs, yet he often went home to the township areas on weekends. Fortunately, he offered to facilitate my entry into the township space. We might consider my fortuitous meeting of Tumelo as the type of interclass (and Afrodiasporic) encounter facilitated by nightlife spaces such as clubs.

A week later, Tumelo introduced me to Thapelo, a nineteen-year-old man who was quite popular in the township and who went to most of the top parties and festivals. I learned early on in our conversations that Thapelo was a huge fan of US hip-hop, so during my interactions with him, I would beguile him with my knowledge of American hip-hop. He was impressed with my extensive collection of hip-hop music, and I offered to make him copies of some of my own CDs for his consumption. He told me that he was a huge fan of Tupac and, in turn, I often included hits and rarities from my extensive Tupac collection in his mixes. We spent time talking about Tupac's lyrics, his life, and why his work moved Thapelo so deeply. As an African American of my generation and as a Californian, the hip-hop movement that included artists such as Tupac heavily influenced me. Thapelo listened closely to my impassioned narratives of Tupac's legacy and importance. Ultimately, our mutual love of hip-hop facilitated an unlikely relationship that led to me gaining trust and acceptance from a group of young men that I would otherwise not have been able to get to know, given our differences in age, sexual orientations, and nationalities. Another thread that facilitated our interaction was the high level of English fluency Thapelo and his friends possessed. Because we were able to communicate on so many levels despite our differences, we were able to share and to bond. I became one of the members of the group. As a result, I was often invited to parties every weekend.

A full inventory of kwaito festival consumption practices is impossible; instead, three of the prominent relationships between self-fashioning and consumption—clothing, cars, and the purchasing of provisions—are highlighted. Clothing is one of the primary ways that individuals ready themselves for participation in these festival spaces. Generally, the very first part of the preparation for any festival is the communication of the body through clothing. Clothing styles in post-apartheid South Africa are heavily influenced by a mixture of Afrodiasporic styles, refigured in and for the local context. I did not, for instance, feel that I had to purchase an entirely new wardrobe in order to fit in once I arrived in South Africa: our Black cosmopolitanisms hung together. In her work regarding Y culture, Sarah Nuttall (2003, 2004) discusses how Y fashion (a fashion connected to kwaito) is created out of a mixture of styles drawn from local African designs and international hip-hop fashion. Loxion Kulca is one company that has fused hip-hop-oriented style with local pantsula style to create a unique South African urban look. Boutique stores such as Sowearto at the Zone Mall in Rosebank have capitalized on this trend. As Nuttall (2003, 2004) notes, labels are important, and popular brands, including Diesel, Nike, Puma, and Levi's, are all represented.

The desirable and sought-after clothing styles of post-apartheid youth tend toward two modes: either imported fashion or prohibitively expensive local boutique lines. Most young people can afford only one or two conspicuous articles of designer clothing. Peer-group relationships, therefore, become extremely important in fashioning the appropriate look for a party. Tailors and designers based in the region specialize in making lookalike clothing that mimics highly desired local and international fashion. Customers clip photos from magazines and give them to these designers so that they can be re-created at a fraction of the cost. Still, such re-creations can easily run well over 600 rand (at the time of research about US$90) for one outfit. Therefore, the ability to use social capital to negotiate installment payments for clothing or the ability to borrow particular items of clothing from friends and family is extremely important. For the average partygoer, a designer ensemble is not simply the private property of the partygoer themself: it is a creative mix of the partygoer's collective resources, both financial and social. The body is central to these constructions. The clothing frames the partygoer's body: it augments and becomes one with it, extending its contours and its boundaries. Fashion is an assemblage of materials, performances, and movement that, together, articulate the self. Looks are scrutinized by peers and outfits are often changed, reconsidered, and changed back again. Yet the desire to look good is not only about framing the body, but also about presenting the body itself. Grooming rituals that involve hairstyling are prominent for both sexes; among women, these grooming practices extend into nail care and makeup application. Hours are spent in salons and barbershops making sure that the right look is achieved. Furthermore, gym memberships, particularly among young men, are increasingly popular: a number of affordable mom-and-pop-type gyms, outfitted with surplus, used equipment, have proliferated throughout the township environments to serve the body-conscious community.

The guys always seemed to want my opinion about a particular shirt, a style of shoe, or a pair of jeans. They wanted to look "sharp" (cool) or "smart" (nice) so that they could attract the best girls at the party. This was, of course, a gendered interaction since I would not have access to young women's preparations for the party, nor would I be privy to what they thought of the ensembles assembled by the young men. Furthermore, despite my queerness, there was an unspoken understanding that their styling was in the service of attracting partners of the opposite sex. Similar to the United States, working-class Black teens were conscientious about luxury brands, and some of these brands (such as Lacoste) were considered special. Thapelo, in particular, was proud of his white Lacoste shoes, which he felt would make him stand out.

I began to learn the language of fashion in the context of kwaito and at the dawn of post-apartheid. I was conscious of the fact that I was older than most of the participants at the festival. Therefore, I made sure that I fit in by wearing brand-name clothing also. I wore a pair of Guess jeans that would stick out, since that style, with unusual zippers and pockets, was not available in South Africa. I completed my outfit with a pair of Puma shoes and a plain, no-name sweatshirt. I understood that my total outfit did not have to be all brand names: one or two well-placed pieces, especially if the pieces were well coordinated, would suffice. I was also aware that, as an African American, I had access to the US market. On many occasions, I could purchase the brands that young people coveted for lower prices than could South Africans (in terms of the US dollar). Since the clothing is imported, and the South African market is relatively small, South Africans had to pay more for clothing from the international market. Not only did I have this advantage, but because I could purchase items that were not available in South Africa, I also had the advantage of owning a unique wardrobe. Uniqueness was also highly prized with respect to clothing. Consequently, it was not uncommon for those who travel abroad to be bombarded with requests to purchase clothing not available in South Africa. One close friend who was a freelance fashion designer and stylist made a significant amount of his income from his ability to travel abroad and then resell clothes at a profit in the South African market.

Making an entrance is an important element of festival participation. After fashion, cars are the second most important medium for kwaito consumption and self-fashioning. The cars with the sleekest look and the most prominent sound system get noticed (see figure 3.3). Cars such as Golf GTIs and entry-level BMWs (1 Series, 325i) and Audis (A1, A3) are associated with the youthful, kwaito vibe. Cars must be fitted with the latest mags (rims) on the wheels, tinted windows, and sport kits. The minibus taxis with the liveliest crowd and the loudest thundering boom also get their fair share of attention: partygoers hang out of the windows with drinks in hand, bouncing rhythmically to the beat. I was often impressed by the ability of my friends to tell me the make and model of various luxury cars. They knew when the latest BMW 700 series was released, and they could identify it. They debated the various merits or demerits of the latest luxury cars with respect to style. Slang terminology has extensive terms for the various different vehicles. The latest-model BMW was known as the BEE (for the Black Economic Empowerment scheme) because of

FIGURE 3.3. Kwaito festival car show participants flaunting sound system, style, and snacks. Source: Pierre Rommelaere.

its popularity among the newly wealthy. In contrast, the Toyota Tazz, a small commuter car available new for less than 100,000 rand, was known as the "starter pack." My own car was woefully inadequate; it was not even a starter pack. Accordingly, our crew made a relatively meek and unostentatious entrance. At one point, one of the partygoers looked at my car and remarked, "This is a no-name car; I can't even tell what it is." In fact, because I was driving such a no-name vehicle, he used it as an opportunity to question my nationality. He was used to seeing the vehicles in the hip-hop music videos and on MTV shows such as *Pimp My Ride*. In turn, he was convinced that no "real Black American" would be seen driving such an old and outdated vehicle. My car did not reflect my African Americanness. I decided to live with this (mis)recognition, understanding that the reality of American Blackness is not reflected in the numerous internationally circulating media cultures that informed South African notions of African Americans' car ownership.

Proper festival preparation means making sure to bring plenty of provisions: the curation and attainment of these provisions is the third practice of consumption that illuminates the performance of self, community, and value that is the kwaito festival. Because these festivals often run until sunrise, attendees must prepare provisions carefully; for our crew, running out of snacks over the course of an evening could be disastrous. Before leaving we would make three stops. The first was a local tavern, where we would pick up alcohol (this would be primarily for my benefit since my companions did not drink). We would then stop near an entirely different tavern to purchase marijuana (this was my group's social lubricant of choice). Our final stop was at the twenty-four-hour convenience store, where we would purchase gum, candy, chips, and soda. Because the vendors at the events themselves would often sell overpriced drinks and food, a well-prepared partier would not be caught without a well-stocked cooler box. Having a well-stocked cooler box is especially important in gaining credibility and respect as a serious partier, as less prepared (or less financially well-off) attendees seek out the better prepared. Negotiations around choosing and procuring provisions, and, in turn, sharing them with peers at the event, impact their status, which is marked by their ability to attract and socialize a quality crew. These negotiations also turn on calculations embedded in economies of desire and pleasure. For example, men can use a well-stocked cooler box to attract women. The same group of men may calculate that it is better to use any extra stash to attract other groups of guys who may then return the favor at a later function. Or, conversely, a group of women may want to be seen at the party with a particularly popular woman. The group may be willing to part with a few of their provisions in order for her to join their group.

As the main social lubricant in these settings, alcohol consumption also marks a form of self-fashioning. Particular ciders are marketed toward women, making their consumption by men unusual or a marker of metrosexuality or queerness. Certain beers connote an air of sophistication or wealth. Given the levels of alcohol consumption in South Africa, alcohol marketers profit from the close relationship between what a person drinks and their self-conception. While advertisers can never guarantee that their product will be taken up in the way that they intend it to, they search constantly for the right combination of "hipness" and popularity among the "cool" crowd in order to drive their sales. University students, targeted by public relations firms, often earn extra funds by serving as brand ambassadors for alcohol companies. In this way, the companies can have their particular brand associated with a specific group of partygoers who possess high social standing, which they hope will, in turn, influence their brand prestige. Public relations firms are also central to identifying those who possess prestige in social circles to invite them into exclusive promotion events that market their brands. The relationship is symbiotic. Preexisting prestige in social circles gets one invited into these "exclusive" party events. Yet being invited to these exclusive party events (often through social networks formed through friends, families, and coworkers) can enhance social prestige.

A well-stocked cooler box can lead to invitations to future events, to new friends, and to additional sexual partners. Usually the negotiations occur between one individual in a group who will approach the group that she or he would like to join. Young men are given more freedom in these situations. They can choose to approach other groups of men that they would like to socialize with, calculating that their access to social capital will allow greater access to additional alcohol, women, or social events. They may also approach attractive women who lack provisions for the evening, calculating the possibilities for sexual exchange into the matrix of decision making. Young women are far less likely to approach men, although they will approach other groups of young women. If they are so bold as to approach men and ask for provisions, both parties understand that a potential, if not guaranteed, sexual exchange may also take place. As young people negotiate the possibility of sexual exchange and its potential range of social benefits, they also understand their relationship to emerging new sexual modes and gender relations. In the socioeconomic shifts of post-apartheid, these politics are at play as more young women have access to cars and careers; the lines of provision negotiation have been blurred.

As Mbembe and colleagues (2008) discuss, in contemporary South Africa, it has become markedly more acceptable for women to go out with groups of female friends unaccompanied by men or boyfriends. Thapelo recounted an incident at the park festival where a sporty car filled with young women approached him. They wanted him to join them for the evening, particularly if he had his own room that they could go to after the festival. It is common in many township spaces for young men to move into free-standing bedrooms, detached from their main house, once they have reached adulthood. Young women are rarely afforded this level of privacy. In the reversal of gender roles Thapelo described, it was the young women who had both the provisions and the car. Yet this interaction also reveals the limits of this role reversal as, despite their economic empowerment, gendered expectations of propriety did not afford these young women the privacy to host men at their homes.

In the scenario described above, Thapelo was propositioned by these young women with the understanding that in exchange for sex and access to his private room, they would offer provisions and transport. Thapelo was unsure, however, whether the women expected him to encourage male friends of his to join this afterparty and thus to "organize" a group of young men for the young women. The English verb "organize" is often used in this context to convey the kinds of exchanges present in these party spaces. Drinks, food, transportation, and potential sexual partners can all be "organized." It might be best to think of organizing as a form of social exchange in which the body and forms of pleasure and self-fashioning are central. While it may be tempting to view such forms of social exchange, particularly when they involve the sexualized body, as merely transactional, scholars should consider that not all forms of capital can be reduced to the economic (see Bourdieu 1986). In an essay on "bisexual" relations in contemporary Côte d'Ivoire, Vinh-Kim Nguyen (2005) works to complicate the assumed transactional nature of sexual relations. Nguyen attends to the relations between openly gay-identified men and their male sexual partners, many of whom identify as straight. He writes, "Homosexual relations could not be reduced to economic strategy, nor were they simply about experimenting with gender roles. Rather, as forms of self-fashioning, they incorporated concerns that were simultaneously those of material and emotional satisfaction, pleasure and desire" (253). Nguyen's observations hold true for forms of heterosexual exchange that occur in party spaces. Furthermore, the types of social exchange inherent at kwaito festivals mean that these spaces allow for a range of nonnormative sexual activities. While the bodies involved in the sexual exchange described by Thapelo were heterosexually oriented, their practices (women as the assertive actors, group sex, and casual sex) were not heteronormative. Lastly, queer people do occupy these festival spaces as well; they are able to use the forms of social exchange outlined here to enhance their social capital and procurement of sexual partners. This queering of sexual interaction occurs despite the overwhelming heterosexual nature of the space.

As the party continued, it became clear that our group's careful preparation had led to a great deal of positive attention and social interaction. All of us, both men and women, were well dressed, which attracted attention from several admirers. Two women came toward our group to comment on Thapelo's shoes and my jeans. Partygoers asked us to share everything from marijuana to alcohol to chips. When we received a stream of engagement and admiring comments, my group knew that we had garnered respect as "serious" partiers. This respect could be leveraged at future parties with the possibility that circumstances might leave us without transportation, funds, or even invitations.

To be recognized as a serious partier, with the correct clothes, the right group of friends, and the right provisions, is to garner serious respect. But it also highlights my point about it being useful to view patterns of consumption in terms of an ethics of the self. Drawing from David Scott's (1999) discussion of self-fashioning in Jamaican dancehall culture, I approach the practices of self-fashioning as practices of freedom. Scott asks us to consider what possibilities are present among those who must make their lives in "the shadows" of dominant power (208). Looking at the Jamaican context, he sees the forms of self-fashioning evident in *ruud bwai* embodiment as practices of freedom precisely because they are "deliberate acts upon the self in an effort to alter the dimensions already imposed upon it, to reconstitute the energies already shaped by existing relations of power" (214). Hence, the practices of Black people in positions of disempowerment might read less as false consciousness and more as self-fashioning.

This self-fashioning is an important facet of the "Black good life." Manthia Diawara (1998) concludes that practices of parading—particularly the masculine parading of self through fashion, cars, and other consumer products—is a form of spectacular self-making: it is about being seen. As George Lipsitz (1994) has noted in his work with African American youth, young Black people who understand themselves to be the object of surveillance both within and outside their communities consciously adorn their bodies for display, with the intent to see and be seen. The Black good life is a "freedom and energy associated with individual fulfillment that had been denied Black people in spite of all the gains in the period of the posts [post–civil rights, post-apartheid, postcolonial]" (Diawara 1998, 242). It is about an unapologetic celebration of life. In the midst of unequal life chances, the insistence on the "Black good life" marks a refusal to reduce one's life experience to the real challenges faced by unequal access to capital in all its guises. My South African friends' obsessive attention to styling in relation to fashion and clothing, transportation, and provision curation does not reveal them to be imitative drones of Western capitalism; instead, they leverage these moments of possibility to articulate themselves through the liberatory aspects of consumption.

The kwaito kids who represent the Y generation have more in mind than individualistic conspicuous consumption. As Caribbean scholar Deborah Thomas notes, these youth "could be refashioning selfhood and reshaping stereotypical assumptions about racial possibilities through—rather than outside capitalism" (2004, 251). Indeed, these images of consumption upset conventional understandings of Blackness, as Blackness is traditionally associated with lack and impoverishment. Even as I locate the liberatory possibilities of consumerism and self-fashioning, I do not discount the detrimental effects of global capitalism on Black South African communities. In fact, I remain sharply critical of these ongoing negative dimensions. It is important, however, to remain critical of the tendency to read the persistence of unequal global capitalist relations in totalizing terms, in ways that further disempower disenfranchised communities by eliding their agency, and their critical practices of curation, self-making, and social exchange. Within consumption and the

Black consumer's desire for "the good life" lies the life-affirming belief among young township dwellers that they too deserve to have a good life. In essence, these young people are saying, "Why can't I drive a nice car, wear nice clothes, and go where I want?" The possibilities inherent in these desires create moments of reconfiguration that have important resonances across and through Afrodiasporic Space. Therefore, consumption patterns are hardly solely conspicuous consumption in the traditional sense; rather, these desires for and performances of displayed materiality connote one possible way that Black youth negotiate histories of consumption in relationship to the multiple ways in which they are disempowered by the contemporary political economy of South Africa. In this way, they remaster the politics of consumption even if only ephemerally by insisting on their right to a good life exemplified through the attainment and display of consumer goods.

The park itself was conveniently located: while in the formerly white areas of town, it existed on the outskirts of the white areas, positioned off a main road that divided these white areas from the township space. Given that this festival was held in the evening, transportation via private car or contracted minibus taxis was necessary, and most attendees arrived through one of these two means. The municipal government sponsored this particular festival, to encourage local youth to vote in the upcoming elections. Thus, it lacked the corporate sponsorship of similar festivals and instead was a space in which the local government promoted its youth programs. In my years of doing research, I observed that this particular type of funding probably structured about 50 percent of the youth-oriented festivals in the early to mid-2000s. Whether sponsoring these festivals had their intended effect of promoting youth civic participation is debatable. The young people who attended the festival, when pressed, did not indicate that they were more likely to vote in the elections or engage in other forms of civic participation as a result of attending these functions. Festivals curated by private promoters typically required admission fees that ranged from the nominal (20 rand) to the more expensive (100–200 rand). Typically, private festivals also had VIP sections whose admission prices included additional amenities (proximity to the stage, screens to see the performers more clearly, food, drink, chairs, shaded or climate-controlled seating) and whose cover charge was typically anywhere from 30 to 100 percent more of the non-VIP price. Those sponsored by the government, such as the one described here, were often better attended because of the lack of admission costs.

The festival grounds were divided into three spaces. The first was the entrance and parking area, which gave everyone the opportunity to see and be seen as attendees arrived in their cars and taxis. This was perhaps the most sonically rich part of the evening, as competing car and taxi stereos battled for attention. In this space, one could get a sense of the latest trends in the music scene, because the more popular songs were blasted by numerous vehicles. Having the loudest sound system and the rowdiest entrance was key to drawing attention to the group. After parking, there was a five- to ten-minute walk to the festival entrance, and this runway formed the second space of the festival grounds. This walk was an opportunity to engage more with the visual and sartorial aspects of the evening. As we walked, we saw familiar and new faces, began chatting with friends, and made plans for the evening. This was also an opportunity to begin preliminary surveillance of one another's outfits, hair, and bodies. The beat of the music that was muffled as one approached the entrance grew louder as we arrived at the chain-link fence that cordoned off the grounds. Once through security and on the other side of the fence, one entered the third space of the festival, the cordoned-off area in which the stage and speakers delineated the official space of the party. While this encounter will focus predominantly on this third space, it is important to remark that the spaces outside the official space are equally key to understanding the social context of the evening. And in some cases, as much partying occurs outside the officially cordoned-off festival space as within it. In addition to marking the mediation between consumption and self-fashioning, festival space offers a window by which we can better understand pleasure, bodies, and dance. It is also a critical element in understanding how the youth remaster subjectivity and space. Festivals offer a window into new constructions of self for Black youth. Furthermore, these parks (particularly those in formerly all-white areas such as the site of this festival) often take on a critically different meaning at night. Inhabited by young Black bodies, these spaces are rewritten away from narratives of (white) suburban domesticity and toward Black joy.

 The emphasis on groups is important, as it organizes the way that bodies move in circulation at the festival. Individuals will leave their groups either in ones or twos to see what is happening around the grounds. I noticed this as my group began to dwindle from five to three and finally to two. Later, it increased to four or five as people who knew us stopped by. A really good song or an anticipated performance by one of the kwaito or hip-hop stars would also stop this movement from group to group. In that moment, the essence of what makes a festival is revealed. The melody of the popular house song began to amplify from the speakers and was followed by a scream of recognition that reverberated throughout the crowd, amplifying through the air. At this moment, the sound from the crowd, an undecipherable cheer of joy, was more significant than the actual music itself. Finally, as the crowd noise died down, people begin the business at hand: serious dancing.

The track's initial melody began in earnest, and the pulsating beat began to take over. Groups of women's bodies began to writhe to the beat of the music, with braids flying from their faces and jackets and jerseys tied around their waists. Groups of men began doing variations of pantsula-style dances. Couples enmeshed as they danced in ways that were beyond sexual suggestion and moved into the realm of sexual simulation. The senses of the body were overloaded at this moment, in some cases enhanced by the proliferation of alcohol, tobacco, and marijuana. The pounding bass of the music that could literally be felt in the body thrilled the ears of the dancers. The sense of touch was stimulated by the autoeroticism of self-touch or by the grinding pelvis of a dance partner. Each dancer appeared out of control of his or her movements but under the power instead of the DJ, who dictated the pace and ferocity of the movements with each selection of a popular track. From house to hip-hop to kwaito, bodies were contorted in a frenzied state of constant joy and arousal. The best dancers occupied center stage, inside circles of movement punctuated by the unregulated flow of dancers from the edge to the center. They demonstrated the latest dances; attendees who were not in the know would become well informed tonight. When it came time for the performances, the crowd turned their bodies and attention to the stage, but the dancing stayed at fever pitch. As the crowd shouted lyrics and focused on the artist, an added dimension of performativity came over the scene.

This encounter brings to light a final conceptual thrust: attention to the cultures of post-apartheid music as they are informed by the features of the field of dance music. Understanding kwaito as dance music links it with notions of the body in ways that expose the pursuit of pleasure. In addition, considering kwaito in terms of its relationship to dance music emphasizes the need to reexamine pleasure as a constitutive category of social analysis. Conventional criticism places artistic production in South Africa, particularly the artistic production of Black youth, primarily in its relationship to notions of liberation and protest. Part of the challenge of post-apartheid cultural production is to explore new and innovative ways of cultural expression. The point is not to dismiss the liberatory or resistant possibilities that emanate from such cultural production; rather, I disrupt the notion that cultural production must necessarily be prescribed by such roles. A careful ethnography of kwaito culture points to the importance of these practices to the politics of self-fashioning that have been a critical platform for Black youth empowerment. Therefore, post-apartheid cultural production in South Africa should be valued beyond its resistant potential. As Saba Mahmood states, "To analyze people's actions in terms of realized or frustrated attempts at social transformation is to necessarily reduce the heterogeneity of life to the rather flat narrative of succumbing to or resisting relations of domination" (2001, 222). While their effects are less readily legible than a clear politics of resistance, I propose a richer frame of analysis, based in a nuanced and embodied engagement with the world of kwaito creativity, that attends to the material weight of these nonbinary creative practices.

In their examination of dance music cultures in Europe, Jeremy Gilbert and Ewan Pearson challenge researchers to consider dance music as an end in itself rather than as "being linked to a safety valve as release for libidinal impulse . . . as expression of emotional hunger" (1999, 15). What they call for is a consideration of pleasure as an end in itself. Dance music and the cultures that both shape and are shaped by it focus on a mode of experience that problematizes the distinction between the inside and the outside. While dancing, the dancer literally feels the music in their body and responds accordingly. This leads to "an experience of music not at all as an object of rational contemplation but as an affect itself whose chief mode of expression is a wordless cheer" (60). Moments of wordless expression create trancelike experiences for dancers, who experience a sense of transportation, and of the sublime. In interviews, partiers have described this alternative place as a sensory one, which produces a kind of euphoria in which they are simultaneously hyperaware of what is happening around them yet completely immersed in the sensory perceptions of their own bodies. Partiers describe feeling both inside and outside their bodies simultaneously.

By the end of the evening, we had been fortunate enough to have several moments of euphoria through dancing. The trancelike state created through these similar experiences of sound *created community*. This festival, in turn, created the space for encounter across gender, ethnic backgrounds, sexuality, and, in my case, nationality. I asked the older members of the crowd to comment on how their partying differed in relation to how the current generation parties. The general response to my inquiry was to link the pursuit of pleasure to the constant sense of violence experienced during apartheid. People needed to experience pleasure in order to withstand the conditions in which they lived and to resist succumbing to the sense that their lives might end at any moment. This was particularly the case during the 1980s, when society seemed on the verge of violent revolution. In addition, they claimed that when they were younger, a party-hard attitude punctuated life experiences during the late apartheid years, precisely because people felt their lives were constantly endangered. Partying became one space that the apartheid state and elders could not control.

In the post-apartheid period, what makes many of the older people I spoke to uncomfortable is that current parties appear very hedonistic and without any regard to endangered political situations, although I am not sure this is entirely true. First, young people still need a space to be themselves that is not policed by parents. Second, the post-apartheid period creates an additional set of endangered embodiments, particularly as the state has pursued neoliberal development priorities that emphasize market-based solutions to social inequality. Foremost among these endangered embodiments is HIV/AIDS infection. Using the frame of productive excess, pleasure should be disconnected from the social-safety valve and should be considered a category unto itself, that is, critically linked to self-fashioning and identity. As Linda Singer notes, "With pleasure arises questions of entitlement and desert, excess and absence, privilege and priority, authority and resistance. Pleasure is therefore already political and politicized" (1993, 72). Thus, pleasure should be considered an important site of agentive expression, a place for remastery. The younger people whom I interviewed often answered that they party because they can: they have the freedom to go places, to do things, and to have fun. Many people asked why they should not be having fun, arguing that life should be enjoyed, especially while young. Another person answered, "I go to a party to celebrate me . . . my boys . . . my girls. . . . It's about who we are. . . . I mean I wouldn't be me if I wasn't at the tightest parties. I am living . . . yeah, I'm alive, it's where I *feel* the most." This sense of energy, this sense of being alive, is one that can barely be captured in my retelling. My experiences through these encounters taught me that pleasure must be read through the multiple possibilities that are occasioned by its understanding as ameliorative and its understanding as constitutive of life.

Ghetto "Fabulosity" in Afrodiasporic Space:
The Example of Izikhothane

Ghetto fabulousness emerges as a key term in contemporary Afrodiasporic popular culture for its ability to straddle the apparent paradox between a space of poverty and enclosure—"the ghetto"—and the notion of wealth and excess expressed through the visuality of fabulousness. Closely aligned with the rise of party-oriented hip-hop music of the mid- to late 1990s (most particularly the stable of artists associated with Sean Combs's label Bad Boy), ghetto fabulousness grew from a particular aesthetic centered in New York, combining high fashion and expensive tastes in consumer products with pride and celebration of being from the ghetto. Unlike the grittier hip-hop typically associated with New York rappers, artists affiliated with this label refused to confine their consciousness to their neighborhoods and spaces of home. Rather, they celebrated their ability to be mobile. Danyell Smith, former editor of *Vibe* magazine, defines ghetto fabulousness as "buying your way up and out [of the ghetto] even if mentally or physically you still live there ... never quite tacky, ghetto fabulous is loud and wrong" (1996, 50). Importantly, ghetto fabulousness is not about being ashamed of having or retaining aspects of a "ghetto mentality." Instead, such retention is a marker of authenticity and credibility. The deliberately oppositional performance of ghetto fabulousness, the willful retention of values and behaviors that oppose bourgeois middle-class sensibility is also about a savvy negotiation with structures of power. The politics of ghetto fabulousness become apparent through the lens of self-fashioning. Indeed, those from these communities know very well that the economic system is structured to ensure their position at the bottom of it, and to deny them access to spaces of privilege should they manage to overcome the odds and obtain access to capital. Ghetto fabulousness flows from an explicit acknowledgment of the racial and economic hierarchies that keep Blacks "in their place" regardless of economic wealth. The refusal of the ghetto fabulous to perform bourgeois behavior enacts a kind of oppositionality that is less about resistance and more about agency. These are a "camouflaged means of negotiating a cultural alchemy where apparent isolation is transformed into contingently employed tactical maneuvers designed to foster inclusion with more mainstream social bodies no matter how fleeting, fraught, and limited that membership status may be" (C. H. Smith 1997, 348).

Like many terms that originate in African American popular culture, "ghetto fabulous" came to take on a life of its own, circulating throughout youth cultures of the African Diaspora and becoming meaningful in a variety of local contexts. In South Africa, ghetto fabulous was quickly taken up by young people; it held a similar valence in the context of post-apartheid. The unique, intersectional relationship of Blackness to wealth, whiteness, and space in post-apartheid South Africa meant that the term functioned with a more acute sense of irony. Because consumption and freedom are so closely related in this context, the notion of ghetto fabulousness disrupts this connection by revealing that, rather than insisting on middle-class domesticity and respectability as the model for post-apartheid freedom, consumption was directed to aesthetic, future-oriented, and youth-centered creativity. In this way, the youth remastered the performance of consumption, divorcing it from middle-class performativity. Again, Mehita Iqani's (2015) discussion of the Soweto golf club members provides an important foil to the politics of ghetto fabulousness. She documents the conventional model for Black post-apartheid agency: heterosexual, middle-aged Black men with families who are appropriately socialized into forms of middle-class performance. Iqani argues that the men of this class see freedom in both communal and neoliberal individualistic terms, and that the tension between these competing visions of freedom defines the stakes of post-apartheid understandings of how the South African government creates or inhibits what we might call practices of freedom. Practices of freedom are at least partly performed through bodily practices such as those that are connected with participation in the party scene and forms of consumption. If the vision of equality of consumption cannot be disarticulated from understandings of freedom, what does it mean when (young) Black people refusing the performance of middle-class respectability attempt to engage in consumptive practices? Following the arguments of Iqani's subjects, we might imagine that there is no difference in how such consumption is represented in public discourses, and yet a cursory examination of media representations proves otherwise. In this sense then, ghetto fabulousness both confirms the relationship between consumption and freedom while also pushing such reasoning to and beyond its limits.

Ghetto fabulousness remains attached to a sense of street culture and a defiant sense of valuation of township origins. According to Ellapen (2007), the township space has become increasingly valued as the repository of authentic Black culture. The liberatory registers of ghetto fabulousness lay in its contestatory valuation of all things township. In this vein, young Black South Africans mirror other global Afrodiasporic youth: engaging in a practice of "tricking back" on whiteness, conventional modes of capital accumulation, and the stark spatiality that divides townships from city and rural areas. Indeed, the fact that such contestatory valuation of the space where the majority of urban people in South Africa reside is necessary, points to the residual traces of colonial and apartheid hierarchies that still exist in South Africa. Kwaito singer Zola uses his seminal song "Ghetto Fabulous" to mark himself as being proudly from the township and to detail not only the challenges of township life but its pleasures, therefore destabilizing the notion that one must leave the township (either physically or mentally or both) in order to experience fabulousness. "Thina ekasi, si-ghetto fabulous [In the hood, we're ghetto fabulous]."

Ukukhothana refers to a collection of aesthetic practices that township youth refer to as *izikhothane*, *s'khothane*, and *ub'khothane*, which in 2012 became the subject of significant media attention in the wake of a number of media exposés concerning what was presented as a subculture of excess among Black poor and working-class township youth. *Isikhothane* (usually written as s'khothane) is the singular and *izikhothane* the plural of those who khothana (show off or boast), describing the person who engages the practice. *Ubukhothana* might be translated loosely to mean showmanship or the state of being a show-off. As slang terminology, the meaning and origin of the word *ukukhothana* is contested, but it might be translated loosely as to boast or to show off. Ukukhothana is an example of ghetto fabulousness that brings into focus contemporary debates about consumption and young Black bodies. Izikhothane are part of a larger conversation about self-fashioning, performance, pleasure, and consumption, particularly as they relate to the bodies of young Black township-based youth. First, by looking at izikhothane as performers, I place them into a longer social history of Black South African performance. Therefore, ukukhothana is a performance practice that shares continuities with historical performance cultures. Second, by questioning the dominant media representations of izikhothane as a practice of wastefulness, izikhothane performance is connected to the concerns that animate this chapter, mainly the reevaluation of consumptive practices attached to "the excluded," the role of consumption in self-fashioning, the politics of exuberance and pleasure, and attempts by Black youth to remaster the signs and symbols of wealth (see figures 3.4 and 3.5).

Some commentary has been instrumental in contextualizing s'khothane culture into a larger sociohistorical frame. In doing so, these commentators reveal the connections between contemporary cultural practices associated with youth and historical ones still practiced by performers that occupy different social locations in urban space. The connections between these cultures speak to the forms of cultural creativity that continue to emerge from Black spaces. In addition, other commentators have reconceptualized s'khothane culture through the lens of performance art. While this commentary has the limitations of reinforcing the category of performance art as "high art" worthy of value and distinct from "street art" and quotidian practices, it nevertheless recognizes the value of s'khothane as a performance.

Perhaps the most obvious example of a similar performance culture that emerges through the lens of social history (and still continues today) is the Swenka culture that developed out of the same-sex hostels in Gauteng. According to Nkosi (2011), "In the 1950s a similar trend arose amongst migrant workers and mine laborers who were subject to the cramped and confined conditions of hostel living. Men, separated from their families and forced into a perfunctory sense of congeniality, would hold contests in which they would

FIGURE 3.4. Young adults with izikhothane, proudly flaunting their status of style and wealth. Image taken at Thokoza Park, Soweto. Source: Amogelang Maluleka.

trade their grimy overalls for the finest suits and flashy two-toned brogues. Called oSwenka, the winner would receive a goat or blankets or maybe some extra money to send home to their families in the Bantustans." Nkosi demonstrates how masculine cultures of fashion and performance are not new elements post-apartheid but are instead deeply rooted in practices of self-fashioning in which migrant workers refused to be confined to apartheid and neoliberal logics. Furthermore, while not explicit, Nkosi reveals the gendered labor of these performances. Much like migrant labor itself, the Swenka performances allow men to reinforce and perform their heteropatriarchal roles in the rural areas through an infusion of cash or material goods needed to support families or raise money for marriage. Nxedlana (2012) also makes the connection between Swenka culture and s'khothane: "To put this current phenomenon in more context you only have to look back to the Swenkas, a group of

"Si-Ghetto Fabulous" 115

FIGURE 3.5. S'khothane display of style wearing the local Ama Kip Kip brand. Source: Penelope Motaung.

working-class Zulu men who took part in amateur competitions that were part fashion show, part choreography, with the purpose of 'displaying one's style and sense of attitude.' . . . Like Swenking, Ukukhothana is competitive. It is a spectacle involving performance and dance, and in both cultures flashy clothing is one of the main symbols of distinction."

Nxedlana (2012) emphasizes continuity between the practices of the Swenkas and the izikhothane: both involve a connection between the use of fashion as a form of social signification combined with competitive performance. Percy Mabandu (2012) suggests that the social history of s'khothane extends beyond the migrant worker cultures of the Swenkas. Mabandu argues that the "need to be seen, this ostentatious posturing by the underprivileged has a long history in the urban Black experience. . . . Earlier examples of showy dressing can be found among the Sophiatown gangs of the day. Going by names like the Americans and the Russians, these ruffians were noted for their sharp attire amid the squalor of that township. The 1970s ushered in a new type of thuggish peacock. Known as mapantsula, they were a throwback to the gangsters of the 1950s." Mabandu suggests how similar performances of what we might call competitive dressing were on display from at least the time period of 1950s Sophiatown and also included the 1970s and 1980s pantsula culture (and their rivals, the Ivies, a subculture that Andrew Tucker [2009] suggests was a noted space of queer male sociality). Mabandu (2012) also reflects on how early kwaito culture involved prominent displays of wealth: "DJ Oscar Warona, for instance, was known for splashing out with ice buckets full of bourbon, which he passed around in celebration. It was the spirit of a country learning how to be free of apartheid's tyranny and defiantly partying it up as if they were already living in a country free of poverty." Thus, a link is provided that reveals s'khothane culture not to be a "new" phenomenon at all but to be a part of a larger social history in which Black men in particular have used forms of consumptive display and clothing to create alternative understandings of their racialized bodies. Monica L. Miller speaks about the historical and Afrodiasporic roots of Black men's fashion, recalling how Black men have historically used adornment to "[convert] absence into presence through self-display" (2009, 10). Through a discussion of the social history of s'khothane culture the various authors remind their readers that displays of materiality and consumptive practices are not new post-apartheid phenomena but instead are embedded in the ways in which Black South Africans have negotiated capitalism.

The idea that s'khothane culture represents a form of performance is an important one. Both Nxedlana (2012) and Boikanyo (2013) suggest that it is more conceptually useful to think of s'khothane battles as a sophisticated form of performance art. Indeed, several s'khothane practitioners have conceptualized their battles as performance, evidenced by the Nkosi (2011) article. While many aspects might be considered performative, three characteristics of s'khothane battles are prominent. The first is the intense preparation that goes into any particular battle. As Boikanyo (2013, 26) attests, "They [izikhothane] spend months planning their outfits and rehearsing their dance moves while they wait for bragging battles that usually take place in parks or other public spaces." The planning of the outfits and rehearsal of dance moves speak to the ritualistic aspect of the battle and to the idea that such battles represent a form of craft that requires training. Second, the competitive dancing and creative wordplay of s'khothane battles represent a combination of "practiced routine and spontaneity" (Nxedlana 2012). Central to the performance is the creative wordplay in which how you say things are as important as what you say. A typical dis battle might involve the following words:

> Oya fosta name ngiya afforda and ngizo ku exposa ngoba ngi cheeseboy ye mpela. Ngiya afforda ukuthenga noku dlala nge cheese. So ungangi tsheli nge Murachinni ngiyi gqoka njenge vest. i-Carvella yi entry level. Ngilala nge Nike Sportswear [You are forcing (trying too hard), I can afford (the most expensive things) and I am going to expose you because I am a real cheeseboy. I can afford and play with cheese. So don't tell me about Murachinni, I wear it as a vest. Carvela is entry level. I sleep in Nike sportswear]. (Boikanyo 2013, 29)

After a turn, the boasting individual will display his items, for it is not simply enough to claim to possess the luxury brands, the brands must be placed on display. The display is twofold, both on the body of the performer who will gesture to the various brands that he is wearing and as he pulls out other such items (expensive clothing the most popular) from a bag or backpack. The ritualized and performative aspect is revealed in what typically happens next. To prove that money is no object, these items are typically rendered useless. Shirts are torn, holes are made in pants, expensive food and pricey alcohol are dropped onto the ground, and the whole pile, discarded food and alcohol, torn and shredded expensive clothing, even high-denomination rand bills, are burned. Part of the performance requires the s'khothane to be completely nonchalant about what would be considered in consumer capitalist logic an extreme and illogical form of waste. Lastly, the performative aspect of s'khothane battles is cemented by their public nature. Local parks in or near township spaces (often the same parks used for kwaito festivals) provide the runway for the izikhothane. The public, often other young people in the community, engages in forms of call and response at each aspect of the performance, urging the s'khothane to create cleverer verbal disses and more spectacular innovative dance moves.

What detractors (see "Izikhothane," parts I–III) of ukukhothana overlook is that s'khothane is not only about dressing the most expensively. It is about dressing expensively with creativity and combining sartorial excess with dexterous wordplay and athletic dance moves. The audience is the ultimate judge of these competitions, and the most famous s'khothane crews can expect that their performances will go viral. However, these viral performances bypass the internet (there are actually few s'khothane performances online). Instead, shared on smartphones via Bluetooth, s'khothane performances circulate through township-based digital media that are inaccessible to outsiders. Therefore, it is not simply winning the battle in real time, but the circulation of the videos and commentary produced from these new media forms that confers social status upon s'khothane crews in the form of shared videos and shout-outs on local radio stations.

Much of the media attention directed toward s'khothane culture suggests that young women are peripheral or absent from the culture; in fact these practices arise largely from performances of masculinity that exclude women's participation. Certainly, gendered expectations would make it less likely for young women to have access to the forms of capital that would be needed to sustain s'khothane performances. There are examples of young women participating in s'khothane culture, particularly its sartorial aspects, but it does not appear to be the case that young women engage in the type of fierce performative battles that are the hallmark of s'khothane culture. Young women, however, are extremely important to s'khothane battles as participant observers who encourage, celebrate, and comment on the performers. In this sense they are as much a part of the performance as the izikhothane themselves. There are also other ways in which the izikhothane and their performances disturb heteronormative conceptualizations of male bodies. Much like other Black dandies in Afrodiasporic Space, s'khothane are figures that through performance become "queer subjects who deconstruct limiting binaries in the service of transforming how one conceives identity formations" (Miller 2009, 11). They reveal the very constructed nature of racial, class, gendered, and sexual identity, placing into crisis the perceived attachment of young township (male) bodies with poverty, hypermasculinity, and heteronormativity.

The soundtrack that knits these performances together is a mix of kwaito, its descendant culture, g'qom, and house music, with a slight preference for locally produced house music. Kwaito and house are more appropriate for s'khothane battles than hip-hop because the latter demands more extensive lyricism and, in some cases, denser sonic production. Because dancing is so key to the portion of the battles that rely on music, performers prefer house and kwaito songs without lyrics (or with sparse lyricism). These soundtracks often emphasize the beats along with a repetitive melodic riff that are sparsely produced, sounding almost like house-oriented break-beats: room for improvisation and for self-articulation between the beats. Sonic space allows the performers to emphasize their dance moves, which often involve complex foot movements that work within and through the beat. Sound in s'khothane performance, however, is often less about the music being played and as much about the shouts, whoops, and hollers of the audience participating and using their voices to signal approval, and the verbal dexterity of the performers who must creatively use urban vernacular languages to their advantage to best their competition. Dancing, then, is simply one component of the competition, although it is the one component that perhaps is most visible for public consumption and adoption. Thus, izikhothane are adept at both learning the latest dances and deploying them creatively in performance, as well as inventing dance moves that then filter to the larger public.

Consumption Matters

Recent discussions of post-apartheid consumption patterns, particularly those of the Black elite and new Black middle class, are instructive for considering the polyvalence of post-apartheid consumer politics. Key studies of consumption patterns in Black communities suggest that these practices are far more complex than the designations of "new" or "conspicuous" placed on them in popular discourses. Several studies (Iqani 2015; Posel 2010) have pointed to the historical relation between numerous colonial-apartheid sumptuary laws and practices of consumption developed in Black communities to thwart these laws. This has meant that consumption has always played a role in the ways in which Black South Africans might configure freedom. More detailed recent studies have revealed that much of the consumption on the part of the black middle class is hardly conspicuous. Instead, this consumption represents the acquiring of goods and services that mark an increase in disposable income and thus has been labeled as compensatory: purchases of satellite television, a car, or new kitchen appliances are all examples of the kinds of compensatory consumption of the new black middle classes.

Through his analysis of s'khothane culture and his interviews with izikhothane, Mnisi (2015, 344) concludes that these consumption patterns are part of identity formation, as they "represent their aspirations to better their situation in life, to escape poverty and the conditions of deprivation, and ultimately become successful." These are also practices of self-formation that, through the regimes of the body, mark a political ethic of the self. That this self-making unfolds alongside and within capitalism does not make it any less noteworthy. Even as we can critique the obvious limits of these endeavors, which do not seek to overturn neoliberal capital so much as re-produce it, we can also see within these practices yearnings for and articulations of freedom: this productive deviance that is embedded in ghetto fabulosity. For these reasons, the izikhothane view their consumption practices as socially enabling (Mnisi 2015, 350), and while the young people I partied with in the park might not have framed their consumption using the exact same language, their understanding of their myriad practices of the self (of which consumption practices are key) are also a form of socially enabling behavior and practices of freedom.

Through looking at consumption, I have found one strategy that those excluded from the markers of wealth and human dignity pursue in order to regain their access to the rights of citizenship. While we can lament the overbearing presence of the market in the lives of Black South Africans, we must also concede that any discussion of human dignity and citizenship cannot occur completely divorced from the market. The young generation of kwaito understands this, so hopefully we take the time to listen to, rather than dismiss, the ways they engage contemporary consumer culture. Kwaito reveals a vibrant, uniquely South African Black performance culture whose consumptive practices reveal possibilities for remastering forms of subjectivity. Through their consumptive habits, young Black South Africans seek to gain mastery over the market forces that pervade their lives as well as remaster notions of Black bodies as pain, and blackness as poverty. The kwaito body reveals a Black body intimately connected with the politics of excess that manifests itself as exuberance and joy.

4

The Kwaito Feminine

LEBO MATHOSA AS A "DANGEROUS WOMAN"

It was a warm and beautiful fall day in North Carolina when I heard of the passing of Lebo Mathosa. I was phoning a friend, as was customary on October 23, to wish him a happy birthday. As I was on the phone, I could hear the music of Mathosa as well as muffled conversations that signaled a festive gathering was occurring.

"Happy birthday!" I said in a joyous ringing tone.

"Thanks. Did you hear Lebo Mathosa passed away?" he responded.

"What!" I exclaimed, shocked at the news and saddened at the thought that someone so young (Mathosa had not yet turned thirty) could be gone.

My friend explained to me how when he woke up in the morning he heard the news that a local woman celebrity, aged twenty-nine, whose name had still not been released, had been involved in a car accident and had died. He explained to me that he did not at first associate the death with Mathosa, given that he had not realized she was nearly thirty, perhaps forgetting that Mathosa had been in the public limelight for nearly thirteen years. When he learned of her death, as a huge fan he was saddened. I asked him what if any effect Mathosa's death might have on his birthday celebrations. "Lebo lived life, so I will live mine. I will celebrate. The party as you hear goes on."

My interest in examining Mathosa specifically, and Black women's bodies and their relationship to kwaito more generally, is to counter the ways that gendered performance within kwaito has been assumed to be solely masculine and attached to cisgender male bodies. If kwaito has been central to giving post-apartheid Black youth a voice, then it is imperative to revisit the performances of the women who were formative in this culture. Kwaito created a space for unfiltered Black youth expression, and while there were clearly gendered and sexual limitations to these expressions, Mathosa pushed against the sexism embedded within kwaito, the larger South African music industry, and South African society to create a space for imagining gender and sexuality outside of societal norms. I consider how Lebo Mathosa strategically deployed the aesthetics of kwaito to become a "dangerous woman" by challenging popular conventions about appropriate femininity through both her onstage performances and her offstage persona. In a context of normative sexualized violence against Black women, her contraventions constitute a dangerous performance because of the possibilities she provides for reimagining Black women's bodies in not only an agentive manner but a "dangerous" one. Thus, her performance practices helped to establish a type of subjecthood for Black South African women that challenged and moved beyond discourses and practices that sought to frame them as passive objects within a heteropatriarchal social field. Ultimately, Mathosa remastered freedom by disidentifying with the "bad girl" construct of Black womanness proffered by Brenda Fassie, the queen of 1980s and 1990s South African township pop.

My use of the term "dangerous woman" is developed through the concept of the sensual-sexual and José Muñoz's (1999) concept of disidentification to counteract the tendency to read all sexual performances of Black women through the lens of hypersexuality. I combine an analysis of her concert and club performances, a reading of her music video "Awudede/Dangerous," and details from numerous interviews with Mathosa in South African media to show how she appropriated kwaito to assist her in freeing Black South African women from the limitations imposed by broader discourses of (Black) gendered propriety. Mathosa's performing Black body is placed alongside other Black women in Afrodiasporic Space to examine how the set of strategies she employed connects her to larger regimes of global Black women's performance even as she accesses and recontextualizes local styles of Black women's performance. The queerness of Lebo Mathosa's performance is a critical part of her dangerous performance. Her queerness is produced through her identification as a queer woman, through her performance strategies, and through the ways that her queer fans appropriate her brand of femininity. I conclude by thinking about her legacy in relationship to Black women's sexuality, considering how contemporary performances by artists such as Babes Wodumo build on the example set by Mathosa.

A Bad Girl, a Dangerous Woman, and the Politics of Disidentification

There is a fine line between sexy and raunchy and I think Lebo crossed the line with her performance yesterday. It was just disgusting. — Anonymous caller to YFM, June 17, 2005

I'm a woman who's living dangerous
The way I sing
The way I dance
It's very dangerous — LEBO MATHOSA, "Dangerous"

They called me controversial when I first came out, and ten years later I remain controversial. — LEBO MATHOSA, accepting her 2005 South African Music Award

These quotes exemplify much about what made kwaito artist Lebo Mathosa a controversial figure in the post-apartheid public sphere. Mathosa, much like Brenda Fassie, had the ability to command attention with the combination of her talent and her capacity to shock. In a 2004 article about Fassie, the South African feminist scholar Pumla Dineo Gqola quoted Gerry Rantseli Elsdon (then Gerry Williams) as saying, "This country needs a bad girl, and [Brenda Fassie is] a lekker bad girl" (Williams, in Gqola 2004, 139). Gqola argues that Fassie's performance of the bad girl created a space for an enactment that "centers women not as objects of some gaze, but as synthesizers of their self-representations" (2004, 142). Gqola suggests that an interesting comparison could be made between Fassie on the one hand and Mathosa on the other. Such a comparison reveals how Mathosa disidentified from the position of the bad girl, "the new Brenda Fassie," and instead created the notion of the "dangerous woman." In doing so, she did not reject Fassie so much as she rejected what Fassie came to represent in the public sphere. Central to my examination will be an analysis of what the performance of the dangerous woman accomplishes in South Africa and why South Africa might need a dangerous woman, keeping in mind that such a performance is already bound to racial representation and material existence.

Mathosa began her career in the early 1990s dancing for a number of local groups in the East Rand township of Daveyton. At this time the neighborhood of Hillbrow was a melting pot for the diverse segments of South Africa's population. Influx control laws were relaxed, and a multiracial, multiclassed group of people began to locate in the city of Johannesburg. Part of Hillbrow's appeal was its burgeoning club and performance scene, where many kwaito artists got their start. Mathosa had begun to distinguish herself in the performance scene of Hillbrow with her athletic and sensual dance moves, stage presence, and powerful voice. Don Laka, a stalwart of the South African jazz scene, was instrumental in the creation of what came to be the first major kwaito group, Boom Shaka. Mathosa was only fifteen when she began performing with Boom Shaka and by the time the group released their first full-length album she and her dance partner, Thembi Seete, were well known for their stage antics and their energetic, sexually charged dancing. Early kwaito performers had to rely extensively on their ability to stage a good performance (or at least a memorable one) and to distinguish themselves from the number of other kwaito groups who were booked for shows. During the early to mid-1990s, kwaito had very little of the exposure that it subsequently enjoyed as part of the music industry of South Africa. Performances at historically Black universities, events staged in community halls, local Black clubs, and stadiums located in township areas were the hallmark of a kwaito performer's publicity. It would not be unusual for a major group like Boom Shaka to book three performances in one day, traveling great distances by bus between performances. Mathosa stood out as the most recognizable member of Boom Shaka for her vocal abilities as well as her dancing style, which tended to gather criticism due to its perceived hypersexuality.

The sensuality and sexuality of Black women's bodies is a charged subject that has a long history in representational and material practices. Here, this genealogy is briefly examined because it frames understandings of Mathosa's public performances of sexuality, particularly the ways in which Mathosa galvanized Black public-sphere discussions.[1] Also, the history of representations of Black women's bodies is intimately connected to the type of cultural labor that Mathosa performs as well as the regimes of power enacted on her body. To begin, Black African women have been historically represented as the height of sexual deviance and abnormality. The hypersexuality of the Black woman's body was (in)famously rendered historically through the body of Sarah Baartman, known derisively as the Hottentot Venus (see Erasmus 2001; Gilman 1985). A second major theme through which Black African women's bodies can be understood is in the discourse of contamination (McClintock 1995). During apartheid, the concern about the presence of African women in urban areas led to them being consistently represented as a threat to the social order, a threat that was a contaminant. A third important aspect that marks the regime of representation of African women's bodies was passivity. Passivity was central to both colonial and local African notions of appropriate femininity (McClintock 1995; Seidman 1993).

Contemporary representations of Black South African women are often complicated by these histories. Writing specifically about the construction of Black women in the post-apartheid moment, Gqola (2007) suggests that one of the central problems confronting the nation is the unwillingness of the post-apartheid public sphere to adequately address the need to dismantle an ideology of militarism that was foundational to the colonial-apartheid state. Instead, this ideology continues unabated in various disciplinary registers. Popular representations emphasize a depoliticized discourse of "women's empowerment," particularly in relationship to the sphere of the workplace (Gqola 2007). However, "outside of work, the dominant gender talk is that women must adhere to very limiting notions of femininity" (2007, 116). These limiting notions are described by Gqola (2007) as a "cult of femininity" that is omnipresent in the public sphere. The cult of femininity is enacted through performances that require women to "exhibit traditionally feminine traits, in other words, that as powerful as women are at work, they submit to the [hetero]patriarchal cult of femininity elsewhere" (116). The message communicated to Black women is that they must modify their behavior, making themselves "seem safe in order to be safe" (121). Black women must "stay at home, participate in the cult of femininity, give in to unwanted sexual advances, surrender many choices, make [themselves] as small, quiet and invisible as possible" (121). The cult of femininity communicates quite "unequivocally that South African public spaces do not belong to the women who live in this country" (121).

While historical legacies and contemporary manifestations shape and contour the regime of representation that Mathosa enters generally, she is more specifically compared to Fassie. In order to understand how Mathosa disidentified from Fassie, it is necessary to review Fassie's performance style. At this juncture, I will not provide an exhaustive deciphering of the complex performance strategies enacted by Fassie over the span of her career.[2] Instead, my concern is with what her performances can be said to have accomplished in relationship to understandings about Black women's bodies. Gqola (2004) argues that Fassie was a transgressive subject, because of her refusal to adhere to the prescriptive norms of acceptable Black femininity. According to Gqola, "Fassie exuded an entitlement to life, to audience, and to a multitude of choices" (2004, 139). Gqola insists that just because subversive Black women's performance could be recouped by heteropatriarchy does not mean that these performances occur outside of agency. Gqola reads Fassie as an agentive subject whose power lay "in complicating the ways in which South Africans talk about sexuality, and women's bodies, [unsettling] the authoritative, limiting depictions [of Black women] which retain currency in public culture" (2004, 145). Fassie's openness about her personal life (particularly her numerous relationships with both men and women) is understood as a direct contradiction to norms suggesting that proper women subjects do not publicly discuss sex and certainly not their own promiscuity. Fassie's constant insistence on pleasure did the work of counteracting the increasing representation of sex and sexuality as it relates to Black women's bodies as the constitutive outside of pleasure. Refusing to be silenced, and rebuffing attempts to portray her as out of control, "Fassie was unapologetic about foregrounding her own pleasure and admiration of her self and body" (144). Fassie took the notion of excess and turned it into a question that problematized the tendency to read all feminine pleasure, particularly sexual pleasure, as deviant within Black women's bodies. As such, she came to represent the "bad girl" and, in playing this role, counteridentified from notions of proper and correct Black femininity.

Bongani Madondo (2014) reveals that part of this counteridentification lay in the ways that Fassie was important as an anti-apartheid figure of resistance, outside conventional definitions of politics and Black femininity, which left Black women as observers to their liberation—important in relation to important men (mothers, daughters, and lovers of the leaders of the nation but never leaders themselves). Fassie's performances were important because they were acts of "renewal and salvation" (Madondo 2014, loc. 151). "With her animated energy, African braids, knee-length shiny plastic boots, sequins galore, the whip-cracking township punk-rock air swirling about her, she gave me and thousands of Black youths the opportunity to see ourselves as we so desired" (loc. 150). Fassie was not an "obedient Black woman," and her fearlessness was part of her critical importance in Black South African memory, creating a space for an alternative politics. "For a while, this girl did more for the proletariat than the whole up and coming union movement could ever dream of at least on the symbolic level. She didn't break as much as dispense with barriers, not only for herself, but for millions of other Black girl tweenies, teens, and young women. Heck, she kicked down barriers for the entire African female and male species" (loc. 150). While not resorting to the bad girl narrative in these quotes, Madondo does suggest the seemingly transgressive nature of Fassie and how her performances created a space for Black youth, particularly young Black women, to reimagine themselves in a time of enormous political uncertainty and instability (see figure 4.1).

FIGURE 4.1. Brenda Fassie. Source: Joe Sefale/*Sunday Times*.

How did Mathosa negotiate both the precedent set by Fassie and the larger nexus of Black women's representation in South Africa to create the performance of the "dangerous woman"? The dangerous woman as a performance resists the tendency to label those that contravene the cult of femininity as "bad girls." If we understand the bad girl as a counteridentified performance, it often exists as the flip side (rather than an escape) from heteropatriarchy, reinforcing its logics even in its rebellion. The "bad girl," for all of her resistant energy, exists as a normative outside available within the logics of heteropatriarchy to instill fear in women who behave outside societal norms of femininity. In the end, we know what happens to bad girls. They typically are punished, denied, as in the case of Fassie, material comfort or steady companionship.[3] Through Mathosa, I investigate the redemptive possibilities of the dangerous woman as an alternative formation to the punitive disciplinary poles of the cult of femininity on the one hand and the bad girl on the other. Mathosa's performances of the dangerous woman were critical for a new generation of

Black women that had to make sense of freedom in explicitly gendered ways that were surprisingly limiting. In what ways might the dangerous woman illuminate alternative possibilities of analyzing Black women's freedom that does not reduce nonnormative women's behavior to being a "bad girl" with all the punitive implications that such a moniker implies? The "dangerous woman" is ultimately a strategic performance that is disidentificatory because it must work through and between normative understandings of Black women's bodies and sexuality, even as it presents the possibilities of alternative formations that are nonnormative.

Muñoz (1999) used "disidentification" to describe the performance politics of queers of color who sought a unique strategy to resist, survive, and challenge hegemonic codes of identity that excluded them. Muñoz suggested that disidentification was an interstitial strategy, one that recognized that a great deal of cultural practices and performances exist outside the resistance/cooptation binary.[4] In the case of kwaito, disidentification is applied to the notion of agency discussed in chapter 3 and therefore disidentification does more than just the work of resistance. Instead, as a performative strategy, it is also a space for contestation linked with the care of the self. To work on oneself is to "veer away from . . . socially prescribed identity narratives [creating] space for new social formations" (Muñoz 1999, 146). In this sense, disidentification becomes a central tenet of an ethic of care for the self as a practice of freedom. Importantly, disidentification is not just about disidentifying from someone/something but is also about a negotiation between identification and subjectification. In Mathosa's case, she negotiated multiple forms of subjectification: one that requires Black women to conform to particular kinds of public behavior on the one hand, while on the other her transgressions are specifically overdetermined by the precedent set by Fassie. Disidentification was a way Mathosa dealt with these multiple forms of subjectification, allowing her to complicate reductive readings of her public persona and to point to alternative social formations of Black women's performance.

Labor is central to understanding kwaito performance. Performance is for these artists literally physical, intellectual, and emotional work; more broadly, it is also symbolic and discursive work. Whether we are talking about the actual physical labor involved in planning, articulating, and performing, or the emotional, psychic, and spiritual labor of performances onstage and off, the different performances and identities enacted all complete some form of cultural work. In his 2013 book *Butch Queens Up in Pumps: Gender, Performance, and Ballroom Culture in Detroit*, Marlon M. Bailey notes the important role of labor with respect to performance in Afrodiasporic traditions. He suggests that one important aspect of the labor of performance is its ability to create new worldviews, an idea expressed through the concept of world-making developed by Muñoz (1999, 18–19). Performances "have the ability to establish alternative views of the world" and in so doing "function as critiques of oppressive regimes of truth" (Muñoz 1999, 195–196). Bailey (2013) and E. Patrick Johnson (2003, 2006) show how the creation of a different kind of space is instrumental to the performance cultures of Afrodiasporic peoples. Furthermore, performance styles embedded in racialized and gendered performativity function as an epistemology, a way of knowing (Bailey 2005; Johnson 2006). Even those performances that might make the critic particularly uncomfortable have value in that they perform the labor of world-making. Therefore, all performances create an ethics of knowledge that intimately connects particular ways of knowing and experiencing the world. This ethics of knowledge is about the valorization of particular knowledges connected to kwaito. Within kwaito, these knowledges create understandings of authenticity and define musicianship.[5] Performing the latest dances, saying the newest S'camto terms, and wearing the latest fashions all mark valued knowledge within kwaito culture.

Performing the "Dangerous Woman"

The scene at the annual Rand Show provided the space for the gathering of a diverse set of audiences. Inside the gallery, older, predominantly middle-class South Africans of various hues perused the different exhibitions. Most of the items for sale included high-priced luxuries, such as furniture, sewing machines, and kitchen appliances. Outside the main venue, there was a different crowd. Occupying a field that was centered by a large stage were crowds of young Black people. Many of those present were from the neighboring townships of Soweto. Some had walked the five kilometers from the township to the setting of the concert, and more would walk back home at its conclusion. Today was a special day for many present, for it marked the end of school vacation. Many in the crowd were students who would be back in school uniforms come Monday. Sunday, however, was a different story.

Despite the torrential rain and the unusually cold fall evening that accompanied it, the crowds were not lacking in either enthusiasm or support for their favorite stars. As the rain continued to pelt the audience, the scene became reminiscent of other such parklike festivals held in inclement weather. Young men began to strip their shirts off and began diving into the puddles of water and mud in the field, almost celebrating the power of nature. Most of the crowd (those who were not lucky enough to be ensconced under the VIP tents) huddled together under umbrellas, the condensation of their breath visible to those around them. Noses ran, teeth chattered, and bodies shivered. The performance schedule was running late, so late that by the time Mathosa was ready to perform, the sun had set. No one had moved, for her live performances were legendary, and she had several big hits to sing and promote. Lebo Mathosa, the former member of Boom Shaka, the controversial queen of kwaito dance, had the power to command her subjects to wait for her performance. And once she was on the stage, her fans were enthralled. People screamed with her songs, they surged to the stage to get a closer look, they danced with abandon, exhausted as much from the act of co-performance as from the wait. They began to disperse once they were satisfied that she had concluded her command performance. As she left the stage, people may have been soaked, but for a brief moment they were no longer cold, such was the electricity of Mathosa's performance. Of all of the aspects of Mathosa's performance that have been subject to commentary, her stage antics, particularly her dancing, have caused the most controversy. In what follows, several assumptions that have dominated understandings of Mathosa's performances are challenged. These assumptions center around the radical equivalence of performance with "reality," the belief that Mathosa lacks any agency within her performance, and the understanding that her performances exist solely or even predominantly in the service of heteropatriarchy.

The first of the assumptions that frame evaluations of Mathosa's performance is the argument that what she presents onstage is somehow equivalent with her actual life. Performance is embedded in processes of making meaning that refer to fantasy, utopia, and pleasure. These things are the everyday and then some, including the things that are not the everyday. Reading performance as simply reflective of reality creates dangerous connections between performance and sexual violence. The caller to YFM, quoted earlier in the chapter, referred to one of Mathosa's performances as "disgusting." Apparently, during a Youth Day concert (Youth Day is celebrated every June 16 to commemorate the youth uprisings in Soweto), she proceeded to dance in a sexually suggestive manner with one of the cameramen who was on the stage documenting her performance. According to the caller, Mathosa's performance was dangerous because it occurred in front of children. The caller claimed that such a sexually suggestive act "made the guys horny." Furthermore, she suggested that Mathosa should be held responsible for any rapes perpetuated by men who witnessed this particular performance. Describing a performance during the height of her success with Boom Shaka, in a 2004 interview with Mathosa, Thami Masemola reminisced on old times, reminding the audience of the outrageous stage antics of Mathosa and her Boom Shaka partner, Thembi Seete: "Suddenly a young fellow yanked up from the frenzied masses is on stage, these two half-dressed cuties playing tricks on him, teasing him as they touch his head, his hands, lead him around the stage, run their hands up and down his entire body. 'Why is that not me' ask someone. Damn that Lebo girl can dance" (Masemola, personal communication, 2019).

Commenting on the dancing styles characteristic of the onstage performance of Boom Shaka, the ethnomusicologist Simon Stephens states, "The first time I saw Boomshaka [sic] dance, I began to wonder whether kwaito music was a reflection or a manifestation of sexual crime" (2000, 265). Stephens links the stage dancing of Mathosa, and other women dancers/singers in kwaito, with the gang-rape practice of jackrolling.[6] In these moments, the tendency to equate sexualized performance with sexual violence should be resisted, a theme that dominates readings of Mathosa. Ultimately the idea is that such performances represent Mathosa's own reality and that such representations are harmful to those who witness them. In contesting this understanding of performance, I am not insisting on the complete disconnection of stage performance from lived experience. Instead, more complicated and nuanced readings of what her performances can be said to accomplish, or in this case, what they do not accomplish, are called for. Whatever the effect of Mathosa's performances, they cannot be understood as merely reflective of her personal desires. Nor are they a masochistic reenactment of rape, enabling the space to imagine the sexual violation of Black women's bodies.

An alternative understanding of her performances sees her as constructive of what Muñoz would call a "public ethics of the self" (1999, 151) (see figures 4.2 and 4.3). This public ethics of the self is not necessarily in service of a feminist narrative of overcoming. However, in providing alternative understandings of Black women's bodies, Mathosa enacted a praxis that freed Black women's bodies from conventional understandings of them. Therefore, her performances created possibilities for remastered freedom if we understand conventional representations of Black femininity as oppressive. Performance is the space for the "theatricalization" of the ethics of the self, linking it intimately with ideas about agency. Mathosa herself was very clear about the fact that there was not necessarily an equivalence between what she presented onstage and her own personal life. Speaking to Dali Tambo in 2001 about his now-canceled television program *People of the South*, Mathosa spoke about her disgust at the fact that men felt empowered to approach her for sexual favors as a result of having seen her performances. Mathosa was adamant that her stage persona did not give anyone the right to perceive her as a sexual object,

FIGURE 4.2. Front cover of Lebo Mathosa's album *Dream* (2001). Source: Gallo Record Company.

FIGURE 4.3. Back cover of Lebo Mathosa's album *Dream* (2001). Source: Gallo Record Company.

particularly not men who were old enough to be her father and thus should have had more respect for her. In another instance, Mathosa stated, "Lebo is quiet and fun-loving. I adore chilling at home with my family, my mom; Lebo Mathosa is different from the stage performer and actress. She loves to relax" ("South African Artist"). "In real life I am a self-conscious person but once I get on stage, the Lebo you know disappears and I lose it—and then the crazy 'drama queen' character takes over" (Mathosa, in Molele 2005). Carolyn Cooper (2004) argues that other Afrodiasporic performers share Mathosa's tactic of separation between the performance personality and the actual person. Writing about Lady Saw, the dancehall queen of Jamaica, whose sexualized lyrics and performance have made her a target of criticism, Cooper states that Lady Saw "discounts those critics who naively identify her, Marion Hall, with her stage persona, Lady Saw. She unambiguously declares, 'Lady Saw is an act.' Pure role play. Distinguishing between her job and her identity, she claims a private space that allows her the freedom to escape her public image: 'I'm a nice girl. When I'm working, you know, just love it or excuse it'" (2004, 111). Similarly, Lil' Kim, the hip-hop artist whose sexualized lyrics and performances made her a target of criticism and debate about Black women's bodies in the United States in the late 1990s and early 2000s, deploys a similar strategy when speaking about herself: "Kimberly Jones [Lil' Kim's given name] is a real person to people. Lil' Kim is NOT supposed to be real. I think that's the misconception. But there isn't that much of a difference, because they are all rolled into one. That's how I look at it, you know? Lil' Kim is a part of Kimberly Jones. But one thing that Kimberly Jones doesn't do when she has her offtime is PERFORM!" (Lil' Kim, in D. Brown 2003).

Mathosa's desire to remind her audience that she is a different person from the performer is part of a disidentificatory strategy that serves two purposes. First, it counteracts the impression that her unrestrained stage performances are identical to being out of control in her life. Second, she reminds the viewer that her performance was a creation, enacted at her discretion. In this sense, she performs an alternative self, one available for public consumption while keeping aspects of her private self from view. C. Riley Snorton (2014) suggests that the creation of such alters is a strategy of queering, which occurs in the more recent Afrodiasporic performances of rapper Nicki Minaj. Of course, as Lil' Kim suggests, the actual demarcation between the "real" self and the "performed" self is never quite as distinct as Mathosa makes it seem. Yet Snorton (2014) argues that this quality of indeterminacy is part of the power of the performance, connected as it is to the market imperatives of music celebrity. The tension between what is sincere and what is performed "underscor[es] the futility of attributing identity—as a fixed marker or index of truth to musical artists. . . . [S]tage names, aliases, multiple identities may be making use of queerness to create different (perhaps more) inhabitable worlds" (297). Moreover, Mathosa's strategy to distinguish the stage performance was enacted to disidentify with Fassie's performative, one that rarely (if ever) made any attempt to separate her performance persona from her actual self, one that left little room for privacy. Mathosa was well aware that despite Fassie's constant denials that she was out of control, the media delighted in using her performance style against her, drawing implicit and explicit comparisons between the person and the performer. Mathosa, in disidentifying from this practice, wished to reiterate that she was in control, but she also created a space that Snorton describes as "unreachable" if not entirely private (297). Her performances were conscious manipulations of understandings of Black women's bodies, used to entice and entertain and critique. This is a strategy employed by Black women performers such as Lil' Kim and Lady Saw throughout Afrodiasporic Space. This idea of being in control is centrally linked to the notion of work, labor, and craft.

Mathosa, in ways similar to Lil' Kim and Lady Saw, consciously linked her performance to the notion of work, and in this way forms a conversation with other working, performing women in Afrodiasporic Space. The type of labor she performed is fraught with additional difficulties, because she enters the contested realm of Black performativity and the gendered expectations within it. Therefore she uses the notion of control and performative personality to present herself as a capable and accomplished businesswoman. Commenting on the occasion of her first solo release, Angela Impey noted that "Mathosa managed to independently finance her latest solo CD single *Intro,* and in the process made a major public impression by the way she represented herself as professionally unconstrained of the heavily male-controlled music industry. Implicit in her example is the message that hard work and extreme self-assurance will bring about personal liberation for women in kwaito" (2001, 48) (see figures 4.2 and 4.3).

Mathosa reminded readers in interviews that the music business is precisely that, a business, and she intended to work hard to make sure she succeeded. This is no easy feat in an industry known for taking advantage of Black women performers. During her interview with Dali Tambo, Mathosa emphasized the fact that historically many Black performers in South Africa have been poorly compensated for their performance labor. Speaking about the late Mahlathini, Mathosa remarked that, despite his illustrious career, he died poor. Mathosa noted that joining the industry as a young teenager left her vulnerable to financial exploitation, and despite all of her success she made little money off of the Boom Shaka enterprise. She spoke assertively about the need to control her image and the business decisions that surround her artistic work, pointing out that she did not intend to die penniless. Instead, by forming her own publishing company, she worked to control the rights to her artistic production. Indeed, Mathosa's estate was estimated to be worth one million rand at the time of her death (Mkize 2012). In a 2004 interview with *Drum* magazine on the occasion of the release of her second CD, *Drama Queen*, Mathosa stated, "I can do all sorts of crazy things but I won't go on stage drunk or high. . . . I respect my craft" (Mathosa, in Molele 2005, 7). Mathosa used the notion of work to disidentify with the presumed understanding that her performance must be automatically linked to drug and alcohol abuse.[7]

In 2004, pictures of Mathosa, performing without underwear, appeared on the internet. In relationship to Afrodiasporic Space, such pictures were immediately connected with similar pictures of Lil' Kim. A South African blog dealing with the subject of feminine (in)decency in the entertainment industry placed the picture of Mathosa without panties next to a famous, widely circulated picture of Lil' Kim similarly (un)dressed. According to the commentator, "There is a fine line between being free and lacking decency and people in the entertainment industry are (in my opinion) flirting with this thin line. I have been disappointed by a number of young female artists—particularly because what they do not only affects their album sales but leaves a message (not a pretty one) in men's minds and influences little girls who look up to them" (Sedumedi 2006).

Placing these pictures into the South African context, over the past few years there was much discourse about young Black women who wear miniskirts without underwear, and the inappropriateness of this behavior—both the wearing of miniskirts and the nonwearing of underwear. The idea is that, in revealing their bodies, young Black women are constructed as disreputable, inviting sexual attention and sexual violence.

In October 2004 the tabloid *Daily Sun* reported that taxi drivers (who are generally men) and passengers subjected women wearing miniskirts and G-strings to physical and verbal abuse. This comment from a taxi driver sums up the situation: "I saw a mother and her daughter both in miniskirts last week and I called my colleagues to help me discipline them. It was shocking, the mother is supposed to teach the girl to dress decently but instead she dressed like her. We got in a group and stripped them" (Ahmed 2004, 2). The notion of women being stripped naked and beaten in public as well as the fact that men felt that these women's bodies were public property available for their censure and discipline is disconcerting. The pictures of Mathosa performing without panties enters into this already-established representational field. In a radio interview with YFM, she insisted that the picture was doctored, and that as a woman and a professional she would never go onto the stage unprepared. This would be to disrespect her work.

Commenting on the pictures herself, Mathosa stated, "Obviously, the person who did this loves my pussy. Whoever is responsible must keep on doing his or her job. But I know you love it—you love looking at it. When these things happen you must know that there is something fascinating about you for people to talk about. If they don't talk about you it means you are boring" (Mathosa, in Thepa 2004, 4). Mathosa used the interview to substantiate her control over the performative space; she suggested that even if the picture was of her, it was enticing, and that there was something about her and the performance that proves interesting. In this instance, by vulgarly reminding people that they are looking at a picture of her genitalia, she forced them to admit their desire on the one hand, but on the other, she reminded them that the object of their desire is hers. Commenting several years later about Mathosa's reaction to the pictures, *KasieKulture*, a blog about township-based culture, had this to say: "She [Mathosa] was casual saying that [just because] you saw it does not make it yours, and it does not mean you will get it and it does not make it less private" (2008). The picture or the performance may allow the viewer to possess her momentarily, but at the end of the day her vagina and her sexuality belong to her, to be used as she sees fit. In the words of bell hooks (1996), Mathosa's response to "Whose pussy is this?" would be to state emphatically that it is hers. This notion of Black women's sexual agency is often denied and has influenced some feminist interpretations of Mathosa's performance as conniving with heteropatriarchy and her own subjugation. Instead her performance of sexuality should be understood to do something more divergent than has generally been assumed.

The second assumption that dominates the understanding of Mathosa is that, in performing sexually, she is somehow conniving with hegemonic representations of women. I consider her performances sensual-sexual, rather than hypersexual, challenging along the way proscriptive norms of behavior. Embedded in her performances are an unacknowledged space of same-sex desire and feminine pleasure, feminine pleasure constructed with the knowledge that the heterosexual male subject can look but cannot touch. Part of Mathosa's appeal is also to queer men and women, who appropriate her brand of femininity in their constructions of selfhood. Two central questions abound: is her performance style merely or even predominantly sexual in nature? If so, what does it mean for this sexual performance to be deciphered as producing space for same-sex desire and appropriation (both male and female), and feminine desire and pleasure that is only partially recoupable by heteropatriarchy?

Mathosa was often described in numerous previously cited reports as grinding against male figures in an overtly sexual manner. While this aspect of the performance was important, it was not the totality of her performance style. The majority of the performances by Mathosa did not feature men's bodies as objects of desire or pleasure. When men did appear in the performance, they did so at Mathosa's bidding, and she was in control of the staging of the male body. With a sly smile and a knowing look, she may indeed grind her body against that of a man or pat him down with her hands. However, she was the one taking pleasure in the spectacle. It is the male body that was on display, sized up for the feminine gaze and dismissed when Mathosa was finished with him. The expressions on the men's faces as Mathosa enacted this performative over their bodies register a mix of desire and fear. Desire, because Mathosa was clearly a woman in touch with her sexuality, embodying the sensual pleasure of women; fear, because in this instance men were not in control of either the situation or the woman's body. In fact, during the performance the man could not know exactly what Mathosa would do next. Part of the pleasure in the performance lies both in its unpredictability and in the reversal of traditional roles, as the feminine is assertive. By performing gender roles differently, she reveals their constructed nature. Criticism of Mathosa's gender performance leads to an interrogation of what constitutes acceptable "masculine" behavior, subsequently denaturalizing this behavior. To mark her performance as deviant marks most masculine behavior in a similar vein.

Her video for "Awudede/Dangerous" also featured Mathosa in relationship to masculine pleasure and gaze, and once again Mathosa showed that she was in charge. The first part of the video features Mathosa decked in neotraditional attire on a throne, fanned by her male subjects. Surrounded by female subjects, Mathosa gazes approvingly at her male dancers. Subsequent to this scene, the musical and visual text shift, from "Awudede" to "Dangerous." The dance-kwaito beat is replaced with the rhythm of Jamaican dancehall reggae. The lighting is darker, and Mathosa is clothed in a fluorescent yellow that glows eerily in the blacklighting used for this segment. Rain pours down as she and two female dancers move their bodies energetically. Framed next to Mathosa is the guest artist on the song, the ragga-rapper Jazz. However, unlike the freely mobile Mathosa and her accompanying dancers, Jazz is tied to a post, unable to move. Mathosa grinds against his body in a style that is reminiscent of the Jamaican dancehall wining style, as she sings,

> I got everyting that a man can need,
> I can shake that ting until you drop dead,
> move that ting until you drop dead,
> I can touch you, touch you until you go ah!
> Move you, move you and get to the move,
> the move that will make your body say what!
> The move that will make your body say what!
> (Mathosa, "Dangerous" [2004])

Jazz's entire body is not free: he can only watch, restrained as Mathosa rubs against him. Furthermore, at one point in the video she holds her arm between Jazz's legs, allowing it to swing freely, making a facial expression of delight, over his apparently pendulous endowment. However, his presence is constructed mostly to prove to the audience the veracity of Mathosa's claim to sexual prowess and subversiveness, although Jazz does suggest that he is one of the few men who can handle such a "dangerous woman."[8]

If being Mathosa's man means consenting to be bound (literally) in the carnival of her pleasure, then this is certainly a radically different interpretation of gender relations in post-apartheid South Africa. It seems that perhaps Mathosa was using the performance space to enact a degree of sadomasochistic pleasure after all. However, contrary to Stephens's (2000) assertion, she is hardly the victim here. Instead, both parties are consensual actors in this space, the man agreeing to be tied up, the woman enjoying his submission. Rather than dance enacting a masochistic remembrance of sexual violation, it is instead a performance space providing a venue for women's (sadistic) pleasure. Commenting on the possibilities of sadomasochism in post-apartheid South Africa, John Noyes argues that sadomasochism must be disconnected from sexual violence. "The [sado]masochistic performance is a staged event that eroticizes socially coded power relationships. It uses theatrical strategies to produce fantasies of power and powerlessness" (quote from Noyes 1998, 141). To the extent that we can imagine the restraint of Jazz as an enactment of sadomasochism, it is important to note that sadomasochism "is not an expression of sexual violence" (146), but it "functions as a form of role playing and meta-communication about sexuality and power" (147).

Mathosa's dancing was not only sensual-sexual but also athletic. It required work, planning, and rehearsal. The dance routines were meticulously choreographed, combining elements of modern dance; jazz; traditional South African, West African, and Central African movements; and hip-hop, Latin, and dancehall reggae. Mathosa did not mindlessly perform sexuality in order to shock and titillate. To the contrary, in the performances that I witnessed, her stage performances were carefully crafted. Costumes were chosen to emphasize the flow of the body during the dance routines. In addition, Mathosa was clear about the fact that men were welcome in the performance space at her invitation. During a 2001 performance in Durban, several young men entered into the stage area while she was performing, and after they were removed by security, Mathosa stopped her performance. She communicated to the audience that these particular men's bodies were an unwelcome intrusion and that their presence meant that the performance could not continue precisely because it would be dangerous to do so. At this moment, the risks of being a young Black woman performing sensuality were very real.

While attending different house parties in Durban during 2001, I observed how Black women of the Y generation used the space of the dance floor to reveal their own sexual autonomy. In much of kwaito dancing, circles are formed, and individuals are encouraged to enter into the circle to show their dancing prowess. On that evening, Thebe's song "Bula Boot" (2001) came on during the festivities. Immediately, several of the women began jumping up and dancing. A circle formed, and two women entered it. Each time Thebe yelled out "*Abuti bula boot*" (brother [man], open the boot [trunk of a car]), the girls opened and closed the zippers to their pants. Dancing with each other, they continued to suggest an air of sexuality and athleticism not disconnected from Mathosa's stage performances. Hands were coyly placed on their hips and buttocks, and at times their buttocks protruded as an enticement for those who were in the circle. As they encircled one another in a frenzy of uninhibited fun, they would continually unzip and zip up their pants. Young men who attempted to enter their space were rebuffed. Later in the evening, Ishmael's "Roba Letheka" (2001) came on. This song celebrates the need to "break your hip," meaning to bend over in a sexually suggestive way while dancing. The video features Ishmael with a horde of young women moving and gyrating to the infectious beat of the song. While this song was being played, numerous people entered the circle. On one occasion, a young man entered the circle. Immediately, a woman entered the space and the young man was exhorted by the crowd to *mnike*, give it to her. On this occasion, however, it was he who broke his waist, much in the fashion of the lithe women gyrating through Ishmael's video. Then, much to my shock, another woman came into the picture, pulled the woman off the young man, and then demonstrated to the crowd that she too could roba letheka.

What both of these examples suggest is that women can use the important space of the dance floor to reconfigure aspects of gender relations. In the first case, Thebe is known for producing songs with extremely lewd and suggestive lyrics. In this case, the women interpreted his "bula boot" to mean the opening of the zipper for sexual intercourse. However they did not let any young man open the boot; they opened and closed it themselves while simulating various sexually suggestive moves. Simultaneously, they rejected the advances of the men who were enticed by this display, thereby showing that they could celebrate their sexuality exuberantly while keeping control over who had access to it. The conspicuous display of female sexuality on the dance floor is thus manipulated and controlled by women themselves. To read the second case is a little more difficult; however, we can see how the politics of the dance floor can exhibit how a man can break his hip just as well as a woman, and that women are also allowed to exhibit and pursue objects of desire just as men are. Thus, the flouting of conventional norms of women's subservience is temporarily upset in this space. As Peter Manuel notes, commenting on women dancers in the Caribbean, "The dance floor is the one arena in which sensuality can be celebrated in a controlled and even artistically created context" (1998, 23). "By dancing with each other, selectively choosing male partners, collectively shouting out rewordings of songs, and other social practices, female dancers actively negotiate their position in society, albeit on largely symbolic levels" (25).

I relate these stories of how women used the dance floor as a space for renegotiating positionalities because Mathosa's performances used this very symbolism as a site for public performances of gender relations. Part of what occurs is the creation of an autonomous space of feminine pleasure where men are acknowledged, but are not necessarily the direct object or the reason for the performance. In speaking about Lady Saw, Cooper (2004) argues, "The erotic performance in the dancehall can be recontextualized within a decidedly African diasporic discourse as a manifestation of the spirit of female fertility figures such as the Yoruba Oshun" (103). Cooper explains that embedded in many of these women-centered ritual practices is the celebration of feminine forms, and the power of women's fertility. Such celebrations often take the form of ribald, sexually explicit language and dancing. In her article on women's initiation rites in northern Mozambique, Signe Arnfred argues that while some feminists have tended to construct women's initiation as a site of oppression, she found instead that, for the women involved, "each new celebration of initiation rites is an occasion for fun and games with other women, for women only. It is a time and space for women, where women are masters. At the same time as they, the older women, are teaching the young ones how to behave in society, and how to confront men, they themselves are taking advantage of the opportunity given by the rituals to enjoy themselves in a space where no man has access" (2003, 4).[9]

Speaking of the sexual aspects of the ritual encoded through performance (speech acts, dancing, singing), Arnfred suggests that "the older women's licentious behavior is often very sexual—but this is a sexuality for women only, sexual ways of playing between women, of which I know no counterpart in western Christian culture" (2003, 5). While Mathosa's dancing is not synonymous with women's initiation, what strikes me is the way in which the denouncements of the practices are similar in that they construct the performative practice of women's licentious sexuality as pathological. There is little recognition of the spaces of feminine autonomy and pleasure embedded in these performances.

Many of the critiques of Mathosa's sensual-sexual performative are also inherently heteronormative, for they fail to recognize the multiplicitous desiring subjects who may be the audience. Mathosa herself was very aware of this, and on repeated occasions reminded interviewers that while she appreciated her heterosexual male fans, she also had a lot of gay fans and that they cannot be forgotten in the construction of her appeal. Within Afrodiasporic Space, Lil' Kim often acknowledges that her fans are not solely heterosexual men: "I have to admit, when I first came out guys were loving me. Then all of a sudden it did a total 360. In my first three months on tour, my shows were packed with men. In the next seven or eight months, the whole front row was women. Now it's a mixture, with a lot of my fans coming from the gay community" (Lil' Kim, in Sarko 1999).

Like Lil' Kim, Mathosa is not sensual solely for the consumption and use of heterosexual men, particularly not in the space of performance. The queering of Mathosa occurs on several levels. At the level of the performance, the heterosexual subject cannot necessarily be assumed, as both queer men and queer women may be a part of constituting the meaning of the performance, with their own interpretations of the dances, songs, and attire. As mentioned earlier, aliases and alter egos (key components to queer performance, particularly in drag) were a key component of Mathosa's performative. In addition, while the performance may be open to multiple interpretations and rereadings, it also has within it spaces of obvious woman-to-woman enjoyment and sensuality. Returning momentarily to the Rand Show concert, many of the dance moves exhibited sensuality between the female dancers. Most of Mathosa's performances did not involve dancing with men in the performance space. Instead, athletic, sensual, sweaty dancing between women marked the performative space. This is, however, not the performance of "lesbian" sensuality for a titillating male audience.[10] Rather it speaks to the possibility that the sensual interaction between athletic, toned, Black women's bodies may produce desire that is hardly heteronormative, having little to do with a masculine gaze or appropriation. Similarly, Black drag performers have appropriated Mathosa's songs, stage presence, and movements into their performances, part of a larger practice of queer men's identification with Mathosa and the power that she has with respect to men. Black gay parties and clubs continue to reverberate with Mathosa's music. Queer men freely enjoy her music, but the abandonment with which it is danced to suggests that they also enjoy her performances, as well as her ability to create spaces of freedom for the open, honest expression of sexuality and pleasure.

Finally, any understanding of Mathosa's performance as queer must also be related to her own performative sexuality. In the case of Mathosa, she did, unlike Fassie, shy away from media attention surrounding her personal life. Mathosa was never publicly linked with a romantic partner beyond her bandmate Theo from Boom Shaka, nor did she openly discuss whom she was dating. Her discretion is a performative practice that is revelatory even as it refuses contemporary narratives of celebrity accessibility. Discretion for Black queers can often be an enabling tactic necessary for survival (McCune 2014). Her queerness exists in a nexus between the performance of discretion and confession. After years of public speculation, Mathosa contradicted the popular assumptions that she was lesbian and identified as bisexual. This subject position is a complicated one to decipher because in claiming to be a "double adaptor," Mathosa refused to be contained within either of the dominant understandings of sexuality.[11] Her public confession of bisexuality confirmed that her performances as well as her own body are entitled to multiple meanings and understandings. Drawing from E. Patrick Johnson's theory of "quare studies" (2001), Mathosa's utterance that she is a double adaptor should be considered a performative speech event that reveals her sexuality and her understandings of her queer desire to be firmly rooted in quotidian Black practices and thus connected to other kinds of queer experiences in Afrodiasporic Space. By confirming that she is bisexual, Mathosa, much like the hijras studied by Gayatri Reddy (2005), reveals the fluidity of sexual identity markers that allow her to be both legible as a bisexual and understood in her community as a double adaptor.

Furthermore, we might consider an additional nonnormative reading of Mathosa as a double adaptor that moves beyond its description as a form of sexual practice to think about double adaption as a survival tactic of Black (queer) women. For Mathosa, negotiating the sexism of society and the music industry required chameleon-like performances, doubling back, moments of silence and disappearance, alter egos and stage personalities, and an "unreachable" quality that allowed her to contest her hyper(visibility) in the public eye. If we understand that Black (queer) women must navigate the twin forms of state and societal surveillance, then double adapting might also refer to the ways that Black (queer) women must always take on different performances in different situations; be nonstatic and adaptable; shift personas and performances to ensure safety; have access to state resources; and avoid criminalization. In essence, double adapting might refer to the complex ways that Black (queer) women's labor draws on guile, adaptability, and perseverance. Mathosa spoke about her need to negotiate and navigate the music industry primarily through her need to disappear from the scene for a while and reeducate herself about the exploitative nature of the industry: "I didn't know much about the industry coming into the scene, and I was always driven around like a puppet and just because you enjoyed doing what you were doing you'd go out on that stage and break a leg. So I had to break out a little bit, learn about the industry, look at my mistakes I made before and choose what it is I wanted to do" (Mathosa, in "Below the Belt Lebo Mathosa"). Aimee Meredith Cox (2015) describes this quality among the Black girls and young women she worked with in inner-city Detroit as "shapeshifting." Shapeshifting describes the ways that these Black women navigate processes of state exclusion. She writes, "Shapeshifting describes how young Black women living in the United States engage with, confront, challenge, invert, unsettle, and expose the material impact of systemic oppression" (7).

Just as Reddy (2005) suggests "with respect to sex" that gender is not an immutable matter of self but rather an unknowable and fluid matter of performance, it might be more useful to see Mathosa's public queerness as a matter of unfixed sexual performance. Whatever her actual sexual practices, it is also possible that Mathosa wished to disidentify from a public performance of lesbianism because of its connection to Brenda Fassie. While Fassie often identified as lesbian, it might be more properly understood, given her numerous affairs with men, that her sexual practices were more fluid than what might be suggested by the term "lesbian." However, similar to Andrew Tucker's (2010) findings in Cape Town, in which he found that township-based Black queer men appropriated the term "gay" for a host of nonnormative sexual and gender practices that would not necessarily be considered "gay" in the Western context, Fassie's understanding of "lesbian" as an interpretive frame for her sexuality might have alternative meanings and understandings that cannot be accounted for by Western understandings of the term. Mathosa's vehement denials of lesbianism, particularly her assertions that such claims amount to "bad publicity," cannot be reductively read as internalized homophobia. Instead, they should be understood in the larger set of identificatory and disidentificatory practices of the self in relation to her own desires. In addition, it also does something entirely different for the audience to read her performances, particularly with respect to men, with the understanding that she is bisexual. Understanding her as bisexual retains the possibility of desire for men's bodies that the performance would lose were we to understand her as a lesbian. "Performative bisexuality" retains the tension between men and women as well as the possibilities of transgression within heterosexual relationships. This transgression means something different if Mathosa is a lesbian who does not ever desire men's bodies.

Mathosa often performed a combination of sexual discretion in relationship to her sexuality that was disidentificatory from Fassie's more revelatory performance of sexuality, as well as in contradiction to the prescriptive celebrity culture that requires a set of confessional technologies with respect to sexuality. In a playful interview with the South African drag performance artist Baroness Coral von Reefenhausen, for the Baroness's television show *Below the Belt*, Mathosa demonstrates the queer possibilities of her sexuality through discretionary performance. *Below the Belt* is described as a "cheeky adult variety series . . . [that] takes an irreverent and risqué look at sex, and alternative culture in South Africa. The Baroness tackles subjects including pornography, S&M, swinging, prostitution, traditional African sex therapies, gay nightlife and other topics 'hidden' in the underworlds of Johannesburg and beautiful Cape Town" ("Below the Belt South Africa"). Vignettes exploring urban South Africa's sexual underworld were often juxtaposed with a number of frank interviews with South Africans concerning their sexuality. During the interview, Mathosa spoke about her numerous piercings, indicating the three visible ones (eyebrow, tongue, and belly button) while playfully intimating that she may have a fourth piercing in the labial area. She stated that she likes to show her tongue piercing during performance and that while audience members question what she is doing they definitely "like it." Playfully rolling her tongue in a suggestive manner, she lets the audience know that her piercing is multifunctional: she can talk, eat, and sing with it since, in her own words, "of course, it works everywhere!" Understanding the sexual implications of Mathosa's insistence that the tongue ring can work "everywhere," the Baroness asks Mathosa about whether she enjoys oral sex, to which Mathosa replies, "Of course." Mathosa, looking down and becoming quite coy, demurred in response to whether she preferred oral sex with a circumcised or uncircumcised penis. Instead, she stated, "It [the performance of oral sex] could be with it." When asked if she was referring to performing oral sex on something artificial, Mathosa rolls her eyes and gestures with her hand in the air and tells the Baroness to "think about it . . . just imagine it, what 'it' could be." After Mathosa proffers a smile and a quick chuckle, the Baroness seems to understand exactly what Mathosa is referring to when she states that her piercing could be used productively in the performance of oral sex on "it." The Baroness states, "Yes, well, I think I know what 'it' is, you have signaled in the most delightful way that we should keep that to ourselves." Mathosa's performative gestures allow the audience as well as the Baroness to read between the lines without actually coming out and saying that she likes to perform oral sex on women. Finally, Mathosa concludes the interview by openly speaking about her desire for both men's and women's bodies, cementing the centrality of queer desire to her sexuality. What this reveals is that Mathosa's "coming out" as bisexual was in some ways unnecessary. Perhaps it was necessary to clarify that she was not a lesbian, but in many ways (through her on- and offstage performances) she had already performatively let the public know she was a double adaptor and perhaps this is why her announcement was met with relatively little public discussion.

Mathosa and the Notion of Violable Women's Bodies

Mathosa's sexual performance should be placed into the context of high rates of sexual violence in post-apartheid South Africa. Speaking on the subject in relationship to her performance, she stated, "As you know, South Africa is the country with the highest figures for rape and woman abuse. I mean we have issues like that. . . . Thembi [her fellow Boom Shaka member] and I were the only two girls in kwaito at that time but we kept Boom Shaka in the public eye because of our controversial dancing, the way we dressed, the sexy way of dancing. It was not easy for the public to accept that because I think the elders found it very dangerous because of their past experiences" (Mathosa, in Kaganof 2006b). While Mathosa is not specific about what constituted the danger of her performance, she alludes to the high degrees of sexual violence experienced by Black women and perpetuated by the colonial-apartheid state, the new post-apartheid state, and within Black communities. However, Mathosa through her performances was critical of a discourse that places undue responsibility on Black women to avoid sexual violence.

It would be a misinterpretation to read Mathosa's performance as leading to or being the cause of sexual violence. Such an interpretation is disturbing on two fronts, for it invests Mathosa with hyperagency while simultaneously evacuating all agency from the individual who is the rapist. It corresponds disturbingly well with Gqola's (2007) observations that Black women are policed by expectations that suggest that their behavior, not the normalization of violent masculinity, is responsible for gender-based violence in post-apartheid South Africa. It places almost all the emphasis of the need to end sexual violence on survivors and silences the need to focus on the rapists and the ways in which society has normalized rape (Gqola 2015). In addition, it conceives audiences, especially young audiences as overly impressionable, that they would reenact the performances of Mathosa violently. Yet there was little forcible or violent about Mathosa's onstage performances of sexuality. Such moral panic had little to do with protecting children and more to do with policing Black sexuality, particularly Black women's sexuality. Writing about the television series *Yizo Yizo*, Clive Barnett argues that what was innovative about *Yizo Yizo* was how the program approached its audience not as passive subjects but rather treated "children as competent subjects with highly developed media literacies . . . recast[ing] the dimensions of media publics for all citizens" (2004, 265). Extending Barnett's analysis, kwaito performance suggests that new public subjectivities are enabled when cultural criticism and public policy abandons the paternalistic view of audiences and instead "recognizes the capacities of ordinary people to participate as active citizens" in the construction of meaning (2004, 254). Second, while it is important to recognize the danger that surrounds assertive Black women's sexuality, particularly queer Black women's sexuality, it is also equally dangerous to conceptualize Black women's sexual subjectivity as located solely within the discourse of violence.[12] This is the mistake that I believe Stephens (2000) makes in his observations about the dancing connected to women's bodies in kwaito. Mathosa provided a radically different rendering of Black women's sexuality, and in her unapologetic celebration of pleasure, she created a space for a different ordering of Black women's bodies.

What a symbolic and discursive reconstitution of Black women's bodies could look like is revealed in the numerous retrospectives written by young Black women detailing how and why Mathosa was important to them. Many of these young women wrote vividly about how the gendered norms and expectations of Black femininity were bewildering to them and how Mathosa offered an alternative model. "Growing up one of three young girls in a family full of boys I knew there were certain rules I was supposed to be governed by that didn't apply to them. I was being reared to be a responsible homemaker, a thoughtful, intelligent, decent young lady and above all else a 'respectable woman.' My aunts frowned upon my habit of sitting with my legs up, my constant slouching and my inability to grasp why there were certain things I had to do because of my gender" (Akoonyatse 2016). As Akoonyatse got older, the advice from women in her family became exasperation, for they could not determine if she was just clueless or openly defiant and stubborn, unwilling to follow the unspoken rules: "I visibly showed no interest in learning how to be that [respectable woman]" (Akoonyatse 2016). This led her to feel like she was somehow a failed woman, a problem that could not be solved. Lebo Mathosa (along with Brenda Fassie) provided an alternative femininity that she could access.

> They became role models before I even knew what role models were. . . . Lebo Mathosa was the cool older sister I'd never had. She might have been young enough to buckle under the pressure society placed on young African women to be meek and demure but she didn't and lo and behold, her world didn't come crumbling down. As obscene as some elders felt she was, as provocative as her persona could be, Lebo was still adored by many. Her youth, talent and love for what she did carried her through. Even those who didn't like how she went about things were compelled to respect her. (Akoonyatse 2016)

Here, the author speaks about how Mathosa's persona and public performances, despite the danger of its contravention, did not make Mathosa a social pariah. The fact that she could still exist and be who she was despite the great deal of criticism she engendered was empowering for this author, leading her to conclude that Mathosa was a role model for a subversive kind of Black womanhood. Lebo taught young women a kind of freedom of moving in the social world that was not accessible in daily life, where respectability reigned as the only acceptable Black woman's performance.

"Dancing Like Lebo": Queering the Construction of Post-Apartheid Nationalism

In this section, I analyze Mathosa's appearance in Malegapuru Makgoba's (2005) much-written-about piece, "Wrath of the Dethroned White Males." Makgoba's essay is an example of the kind of dismembering of Black women's bodies that occurs in the service of nationalism in post-apartheid South Africa (Samuelson 2007). Makgoba, who was at the time the vice chancellor of the University of KwaZulu Natal, wrote an editorial in which he lamented the lack of transformation occurring in post-apartheid society. Comparing white South African men to dethroned leaders of a congress of baboons, he suggests that the slow pace of change is exacerbated by white men's unwillingness to respect the authority of the new leaders and rightful heirs of post-apartheid South Africa, presumably Black men. According to Makgoba, this new social order would be decidedly African and that the white South African men who were truly interested in participating in the post-apartheid nation needed to adopt African cultural practices. As an example, Makgoba (2005) stated, "It should therefore become common sense that the white male soon learns to speak, write and spell in an African language; that he, like Johnny Clegg, learns to dance and sing like Ladysmith Black Mambazo. He should learn kwaito, dance like Lebo, dress like Madiba, enjoy eating 'smiley and walkies' and attend 'lekgotla' and socialize at our taverns."

An analysis of the various different responses to Makgoba's essay is beyond the scope of my examination. However, many of these responses revealed the impoverished state of national discussions of heteropatriarchy, its connection with white supremacy, and the continuous ways in which these two forces work together to structure post-apartheid South African society. As Andrea Smith (2006) points out in the United States context, heteropatriarchy was central to the construction of the American nation that relied on the three pillars of white supremacy: slavery/capitalism, genocide/colonialism, and orientalism/war. For Smith, it is important for different people of color in the US context to understand how their bodies might be positioned in relationship to these three pillars in order to recognize the different ways in which a heteropatriarchal white supremacy situates their bodies in relationship to American empire. I find her schemata useful because the construction of modern South Africa relied on positioning Black bodies simultaneously in each of the three pillars that Smith identifies.

Mathosa's dancing body, her performing body, is called upon by Makgoba to do particular discursive labor in the service of his argument. Mathosa as an individual disappears in this formation. Her body and performances become ciphers to build a nation where white men learn to dance like her and Black men use the (in)ability of white men to dance like her as a weapon to empower themselves. Black women and their cultural and creative expression are not useful for what that tells us about Black women, rather Black women's performance is useful in relationship to Black men. As Suzanne Leclerc-Madlala (personal communication, 2010) suggests, there is no room for women in Makgoba's formulation of the post-apartheid state. Rather, women become possessions, like other resources of the state to be fought over by competing racialized patriarchies. In creating an analogy whereby the powerful elite of South Africa is constructed as a congress of primates, whereby one psychoanalytic figure (the white male) is replaced by another (the Black male), Makgoba suggests a form of transformation that can read only (Black) women as spoils for the victors in this hierarchal heteropatriarchy. Mathosa is useful in that her athletic dancing becomes a symbol of lack for the white (male) body. Suffice it to say, such a nationalist rendering of Mathosa is a violent, reductive reading that leaves out her dangerous qualities. That is, if we take Makgoba seriously, that being part of the post-apartheid nation would mean tapping into the erotics of the Black feminine, how might we productively reimagine the post-apartheid society that exists? What might it mean to really dance (or perform) like Lebo Mathosa? What kinds of practices are central to her performance that must be dismembered in order to serve the Black heteropatriarchal, nationalist fantasy?

Her video "Awudede/Dangerous" provides an opportunity to think productively about what dancing like Mathosa might actually mean in the context of post-apartheid South Africa. Mathosa's video is divided into two portions. In the first portion, Mathosa's dancing is to the song "Awudede," in which she portrays a regal figure who could be read as the rain queen of the Balobedu. As I mentioned earlier, she commands her subjects as they dance for her. While she does join the dancing at one point, the performative movement focuses mostly on her male and female dancers, with women surrounding her while men fan her. During this vignette as Mathosa and her subjects dance, not only is she singing in Zulu rather than Khelobedu (the language of the Lobedu people), but neither she nor her dancers limit themselves to the traditional dances of the Balobedu people. Instead, Mathosa and her dancers draw from hip-hop, capoeira, Nguni, and Congolese-influenced popular dances. In the second portion of the video, as rain cascades around her, Mathosa wines against Jazz, referencing the moves made famous by Caribbean dancehall queens. As Gqola asserts, "Young artists working in all genres seemed to intensify the production of a South African aesthetic: *cheeky and unapologetically mixed as it drew inspiration from wherever it saw fit, and so it became 'Afro chic,' more kwaito, more 'Afro pop,' more Mzantsi, and so on*" (Gqola 2007, 112, emphasis mine). It is this mixing, particularly from other Afrodiasporic spaces, that makes Mathosa's performances possible, that makes them so "cheeky," "so unapologetically mixed," and therefore so "Mzantsi." It reveals Lebo's body, the Black South African body, the South African nation to already be formed in relationship to the transnational Black body (see figure 4.4). Furthermore, the video also reveals the lingering feminist potential of this performance. As Meg Samuelson (2007) correctly points out, the persistent (hetero)patriarchy of the post-apartheid nation and its violence require an imagination for Black women that is transnational.

FIGURE 4.4.
Lebo Mathosa.
Source: Instagram:
Lebo Mathosa.

Mathosa's sly reference to the Balobedu rain queen is central to our understanding of the visual and performative text at play here. The ceremonies and dances performed traditionally by the rain queen's subjects reveal the queen to have special powers; she is in control of the clouds and rainfall, so she is in control of life itself since the rain provides sustenance to the people. Historically, her powers were respected by many of the neighboring traditional chieftaincies. Mathosa's reenactment of the rain queen's court is a performance that re-centers feminine power and the sacred possibilities of the erotic. Her female subjects writhe in rhythm with Mathosa's own movements, their bodies inhabiting a woman-centered sensuality. If, traditionally, the rain queen and the dances of her sacred court are expected to guarantee rain and thus life, we can see how Mathosa employs this trope within her video to more explicitly reveal the possibility of the feminine sensual-sexual and the erotic potential within. Mathosa centers her own gaze, and through dance she controls both women's and men's eroticism.

At the conclusion of their dance, the shift in visual and sonic text from "Awudede" to "Dangerous" reveals that the performance has had its intended effect of producing the rain. Mathosa now centers her body in relationship to Jazz's body while drenched in the wetness that she and her subjects have summoned. Through her performance with Jazz, she centers her sexual desirability, implicitly and explicitly linking the sensual-sexual to the life-giving potentialities of the erotic. Using performance, Mathosa queers tradition as well as the South African past through her remembrance of the rain queen's rituals and her use of them to remind us of the possibilities of woman-centered power. Yet there is another queer aspect to Mathosa's choice to perform the rain queen. Mathosa, having grown up partially in Tzaneen, would have been very familiar with the power and potentiality of the rain queen as a way to communicate feminine and queer erotics. The queerness of the performance rests on Mathosa's use of rainmaking as a trope to reference the rain queen and the sensual-sexual, erotic possibilities contained within the practice of rainmaking. Writing about the queer use of rainmaking in Zoe Wicomb's short story "In Search of Tommie," Andrew van der Vlies makes the following observations that are worth quoting at length. He writes,

> The most famous rainmaker in South Africa is undoubtedly the Modjadji, the Rain Queen of the Balopedu people in Limpopo.... What makes these traditional leaders additionally interesting for our purposes is that they do not marry male consorts; they are allowed numerous "wives," women courtiers and assistants whose children are considered hers. The Modjadji is not (necessarily) lesbian-identified, though she is on a continuum with a broader occurrence of lesbian traditional healers, or sangomas, who (particularly in parts of KwaZulu-Natal and the Witwatersrand) commonly take "ancestral wives," responsible for serving "their female sangoma husbands" in forms of "marriage" of great variety: in some, "sexuality is often taboo," although the work of Ruth Morgan and Nkunzi Nkabinde has demonstrated "the existence of secret"—as well as open—"sexuality and sexual relationships among certain same-

sex sangomas." Queer characters, queered performances of faux traditions, cultural practices more open to queerness than Bible-thumping or leopard-skin-clad leaders would have us believe, are everywhere, in other words. (2011, 440)

Much like the interview with the Baroness, Mathosa communicates performatively her queerness through her repurposed use of the rain queen. By centering women's desire, women's authority, and the erotics of sensual sexuality, Mathosa issues a direct challenge to heteropatriarchal authority. If being South African really does mean dancing like Lebo, then we might unleash the productive, political, queer possibilities of post-apartheid South Africa, one that recognizes and empowers the centrality of the erotic, of women's bodies, of transnational imaginaries in the construction of the post-apartheid polity. Given the heteronormativity and heteropatriarchy of the post-apartheid nation, the vision that Mathosa presents of post-apartheid freedom is a needed yet dangerous one.

The combination of the cult of femininity that prescribes Black women's roles and the specter of militarism that disproportionately disciplines Black women's bodies reveals why South Africa needs dangerous women like Mathosa. Certainly dangerous women in other genres of music, like Simphiwe Dana and Thandiswa Mazwai (who got her start in kwaito but now predominantly performs in a variety of other genres), have emerged to challenge limited conceptualizations of Black women's performance. Yet in popular youth genres like kwaito, there seems to have been a significant void left by Mathosa's untimely passing until the recent appearance of Babes Wodumo. Brenda Fassie was certainly a transgressive subject. In fact, using Muñoz's (1999) typology, it might be fair to understand her as a counteridentified subject. The politics of late apartheid South Africa and early post-apartheid South Africa positioned her as such and perhaps necessitated this kind of performance. However, Fassie paid for this transgression with her life, an increasingly self-destructive lifestyle of drug dependency, which ultimately led to an overdose. Mathosa disidentified from the narrative of Fassie and suggests, "My fans don't care who I'm sleeping with or what I do in my spare time, they love me for my work" (Mathosa, in Ngudle 2004, 15). She was uninterested in having aspects of her personal life splashed across the media and suggested that perhaps her fans were as well. At the very least she believed that her sexual partners would not be the reason for her fans' interest in her music. She fought to maintain an element of unknowability. Mathosa, much like similarly positioned women in Afrodiasporic Space, used the notion of labor and work to disidentify from the tendency to equate performance and reality to a simply reflective relationship. By representing herself as a businesswoman, she also highlighted how the dangerous woman is a construction to be used and manipulated as she sees fit. The dangerous woman reminds us specifically of the material realities of Black women's performance.

In sum, Bongani Madondo had this to say about why in the current moment Mathosa's memory is so key: "in this—our Beyoncé feminist times, our Zanele Muholi queer-me-nist times even, our Babes Wodumo its-my-time-and-I-will-have-my-cake-and-eat-it-era, all of which Lebo not only performed but lived and encapsulated, the idea of Lebo Mathosa Matters cannot be overestimated" (Blignaut 2016). Like in this chapter, Madondo places Mathosa into a genealogy of Black women's performance that is simultaneously rooted in South Africa and Afrodiasporic Space. Considering how freedom is remastered, if we take seriously how Fassie was able to empower her audience through representing radically different forms of Black women's performance, then we can understand how Mathosa's contravention of social norms participated in doing something similar for the post-apartheid generation Y. What she meant for Black women and queer folk in general, what she especially meant for Black queer women, was that she offered a way to imagine the Black queer woman's body, the inappropriately feminine body as central to the construction of the new nation. Through her performances, freedom could be remastered for Black/queer/women, shifting freedom away from being embedded in a process of transfer of power and authority from one set of racialized heteropatriarchy to another.

5

The Black Masculine in Kwaito

MANDOZA AND THE LIMITS OF
HYPERMASCULINE PERFORMANCE

On September 18, 2016, the nation of South Africa mourned Mandoza's death. A number of tributes from political figures to entertainment personas emerged in the aftermath of his untimely passing. Most of these tributes focused on Mandoza as a symbol of national unity, a redemptive figure who crossed over and brought whites and Blacks together. The following quote from President Jacob Zuma, released by the Office of the Presidency, highlights this framing of Mandoza's importance to post-apartheid South Africa: "South Africa has lost one of its pioneers whose music appealed to a cross-section of our people, young and old and was known to have achieved the unique crossover culturally to be enjoyed by both Black and white South Africans" (Zuma, in Laccino 2016).

This chapter builds on the analysis of gendered feminine performance offered in chapter 4 in order to explore the possibilities created by the masculine kwaito body. That kwaito reveals but also influences Black masculine gender performance is almost axiomatic: it is a male-dominated art form entrenched in certain forms of masculine gender performance and patriarchy. However, kwaito subjects enact multiple forms and strategies of subversion and reconfiguration to imagine new possibilities of freedom for male bodies and masculine performances.

In this chapter, I focus on how Mandoza's performances enabled him to strategically rescript the image of the thug (tsotsi) in South African social discourse, and how his reception within Black communities shows how his manipulations produced a redemptive and mobile masculinity that could transverse different and sometimes contradictory landscapes. In contrast to much of the analysis of Mandoza that views him as a figure of racial reconciliation, here the intraracial possibilities of his performance take precedence. In transforming the image of the thug, Mandoza collaborates with the rehabilitation project of the Black elite. While the persona of the thug lays outside the expectations of Black bourgeois heteropatriarchy, Mandoza represents a transformation from poverty and waywardness to a respectable family man. The Black (male) bourgeoisie was able to use the lower-classed Black masculine body to articulate their own notions of post-apartheid freedom and masculinity. However, Mandoza understood thug performance as a commodity and used his mastery of the performance to enact his own forms of agency, creating a disidentified "focused thug" persona that challenged commonsense understandings of the tsotsi. This remastered thug performance became a performative masculinity available for desires whose queer implications destabilize the inherent normativity of performative Black masculinity. Ultimately, Mandoza remasters freedom by disidentifying from and reframing Black thug performance, in the process revealing both its productive possibilities and its limitations.

Mandoza emerges as a central figure within kwaito in 1999, with the debut of his first solo album *9 II 5 Zola South*. Similar to Lebo Mathosa, he emerged as a central figure and most compelling member of an already-established kwaito group, Chiskop. Mandoza's performances function in disidentificatory ways with respect to the tsotsi. Particularly within Black communities, complex configurations of desire that center Mandoza predominantly in the specific ways in which he constructs, contests, plays with, and acknowledges what it means to be a tsotsi. Mandoza offers the audience a performative masculinity, one that is already inherently racialized, classed, sexualized, and, with respect to thug identity, in need of rehabilitation. What does it mean to perform the thug, particularly in relation to the constant balance between rehabilitation and exclusion?

While Mandoza was not the only kwaito artist to engage with a tsotsi image, he was perhaps (save Zola) the most well known. Many other kwaito artists

(Mapaputsi, Mzekezeke) perform other types of masculine identities, namely the *kleva*, which may use some of the markings of the tsotsi (i.e., language, attire), but whose image is generally divorced from the explicit criminality and violence associated with tsotsi-ness. The tsotsi is a figure who at various times has inspired a mixture of begrudging admiration as well as fear. Black South Africans recognize that making a living in a system of racialized inequality may mean having to resort to a number of "hustles" that cumulatively cross the lines of legality and morality. According to Cheryl Keyes, in hip-hop, a hero is often reconfigured not as someone who saves someone else, but as "someone who has stood against the odds and adversaries of life" (Keyes, in Neal 2003, 208). Therefore, the thug can in some instances be reconfigured heroically as a survivor. Importantly, this is a gendered notion of heroism, for feminine figures who exist outside of legality and morality rarely receive similar romanticization. An ambivalent relationship with the figure of the tsotsi exists in Black communities. While many admire an ability to hustle and survive by any means, they often disagree with the methods used in these hustles and the tendency of tsotsis to prey on their own communities in addition to wealthy (white) communities.

The tsotsi has his counterparts elsewhere in the Afrodiasporic Space and exists in the circulating anxiety about Black male subjectivity produced within, outside, and between Afrodiasporic communities. In US hip-hop, the association between music and thug personalities was prominently featured in the 1990s debates about West Coast (Los Angeles–based) "gangsta rap." Individuals as diverse as C. Delores Tucker, William Bennett, and Tipper Gore all were part of the moral panic that surrounded the rise in popularity of gangsta rap. While the agendas of the gangsta and the tsotsi were diverse, the gangsta figure as a site of young Black male pathology remained similar. Part of the concern was interracial, given the growing popularity of the music in suburban white America. In addition, there was an intraracial critique that such performative masculinities in music contributed to violence in urban US Black communities. Similarly, in the postcolonial Caribbean, the rude boy and his association with dancehall has been a site of anxiety, particularly as young Black men become increasingly figured as the site of irredeemable pathology. Donna Hope (2006) demonstrates how, lacking the traditional avenues of work necessary for securing patriarchy, young working-class and poor Black men in Jamaica have increasingly turned to crime as a means of survival. The associations between this class of men and dancehall music have made dancehall the site of intense discussions about the postcolonial Jamaican nation-state. In the case of the tsotsi, the concerns about intraracial pathology prominent in US debates about gangsta thugs and gangsta rap combine with concerns about class and nation prominent in debates about the dancehall thug and "slackness." Therefore, I envision that discussions of Mandoza's tsotsi performance will resonate with thug performances as they exist elsewhere in Afrodiasporic Space.

Critical to my analysis is the understanding that the tsotsi as a figure exists apart from kwaito music, yet the association between kwaito music and "the thug" has been prominent from the beginning. Kwaito music trafficked in the slang and argots of township South Africa, which were genealogically connected to criminal gangs. Second, a number of prominent artists in kwaito performed thug personas, their music offering a glimpse into the life of the tsotsi, or a narrative whereby the artist referenced a criminal past that had been left behind. Lastly, kwaito—incubated in the shebeens of the townships and nightclubs of newly integrating Johannesburg—was often connected to a sense of nighttime spaces of leisure being sites of criminality and disrepute. One possible explanation of the word "kwaito" as the name of the musical genre draws its meaning from a well-known gang operating at the time (see Stephens 2000). The symbiotic relationship between the tsotsi and kwaito, particularly in Black masculine performance, is key for understanding how the culture operated and was received, particularly in its early years. Hence, my analysis draws on the tsotsi as a figure that emerges within kwaito but has a larger societal genealogy (see figures 5.1 and 5.2).

The rehabilitation of the figure of the tsotsi into productive pursuits has been a preoccupation of various segments of South African society, both contemporarily and historically. Recent concern about the tsotsi can be seen in the Academy Award–winning film *Tsotsi* (2005), directed by Gavin Hood and based on the novel by Athol Fugard. The problem of the tsotsi is popularly dramatized through the lens of this film. In the film, the tsotsi figure comes to understand and achieve humanity by going through an existential crisis brought on by a carjacking incident in which he unknowingly kidnaps a baby. The film's resolution rests on the image of the criminal justice system as offering restorative justice, which contrasts with the reality of race and policing in post-apartheid South Africa. While the tsotsi figure is able to undergo a process of redemption through self-analysis, the film suggests that this is insufficient. True redemption occurs not at the level of self-realization, or in community, but at the level of punitive state discipline. During apartheid, South Africa ranked number one per capita in the number of its citizens incarcerated. Currently, South Africa still incarcerates a disproportionate share of its citizens, and this incarceration rate is heavily racialized and classed. The tsotsi finds resolution/redemption through surrender to the disciplinary apparatus of the

FIGURE 5.1. The cover of Mandoza's album *9 II 5 Zola South* (2010).
Source: The CCP Record Company.

state (symbolized by a Zulu-speaking white male officer), suggesting that a racially skewed criminal justice system is able to provide redemption for Black men criminalized by the state and society.

In his text *Refashioning Futures: Criticism after Postcoloniality* (1999), David Scott addresses the issue of what it means to critically analyze figures like the tsotsi. Scott suggests that the notion of alienation is inadequate in thinking through the theorization of the body in contemporary postcolonial societies. Traditionally, the figure of the working-class subject is positioned as young, Black, male, urban, and angry—its antisocial behavior a form of resistance "that haunts the middle class imaginary" (1999, 210). This figure is the counteridentified subject, the subject who does not simply lack the traits deemed respectable to the creation of the postcolonial state, but who refuses these

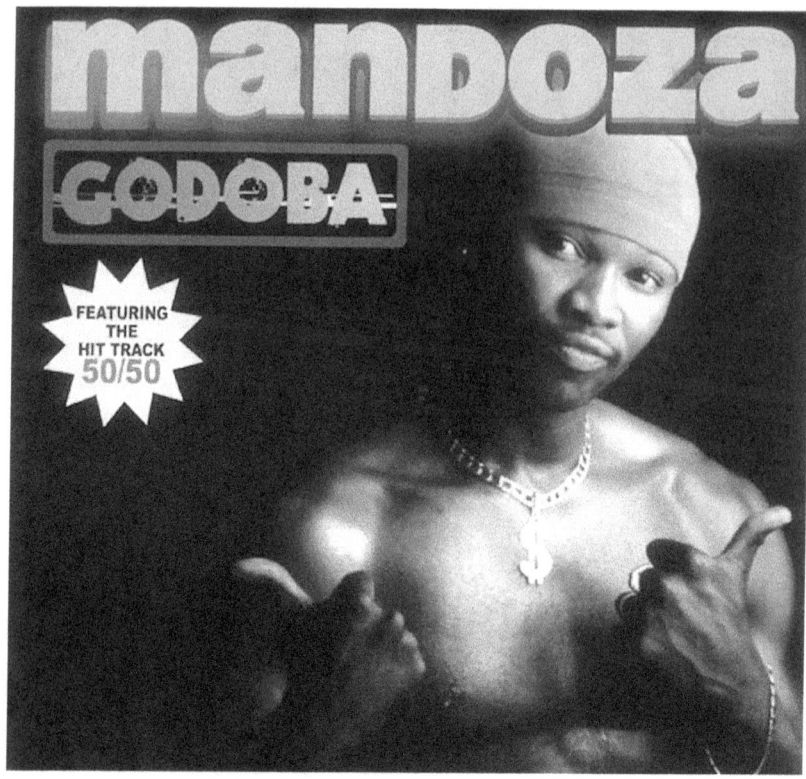

FIGURE 5.2. The cover of Mandoza's album *Godoba* (2001).
Source: The CCP Record Company.

values. The body is repressed underneath a truth that is buried under the weight of colonial power. Subsequently, it is liberated through either violence, proper reeducation, or both. This (always masculine) figure can then be recouped by liberation movements under a banner of revolutionary-liberationist overcoming, from alienation to realization. However, contemporary realities make such a reading difficult to sustain. In the case of South Africa, the redemptive rendering of the tsotsi was popularized in liberation discourse through the figure of the comrade-tsotsi. Through both education and violent resistance to the apartheid state, the comrade-tsotsi represented rehabilitated possibilities as the tsotsi morphed into a comrade. His violent tendencies were reorganized in a Fanonian (1963) turn away from his community and toward the apartheid state as the site of oppression.

Thokozani Xaba (2001) examines precisely what happens to the comrade-tsotsi after apartheid. Short on the skills needed to compete in the post-apartheid economy (having dropped out of school to fight for liberation) and decommissioned from serving in the new multiracial military, these men find that the very masculine values that served them during the late 1970s and 1980s are pathologized as problematic in the post-apartheid order. These men (re)turn to criminality to support themselves, marking their bodies as increasingly unwanted and undesired symbols of post-apartheid masculinity. As masculine, dangerous subjects in the contemporary period, they now appear as little more than unpoliticized criminality. They also become an easy representational trope through which to deny legitimate political protest against the post-apartheid state by conceptualizing such protest as criminal behavior.

Following the insights provided by Scott (1999), it might be more productive to reread tsotsi performances not as repressed internalization of colonial violence but as practices of self-formation. In the case of dancehall music, Scott suggests that the gestures, poses, and behaviors of the rude boy be read as deliberate practices of self-fashioning. These processes of self-fashioning allow the rude boy to construct practices of the self that are understandable as an ethics of self-care (see Foucault 1987). What is dangerous for the postcolonial/post-apartheid state is precisely that these expressions are not easily recoupable, and attempts to do so lead to what Suren Pillay (2008) has identified as forms of critical disappointment. Disidentification works as a way to understand the practices of self-fashioning and care of the self in relationship to these subjects precisely because they are not the counteridentified subject. However, in not being counteridentified this does not mean that they are solely capitulating. Instead of falling into a form of critical nostalgia, what I propose is that the expressions of Black men and masculinity connected with the tsotsi be understood in a nuanced way that recognizes both the limitations of these expressions and their productive possibility. As the discussion of the comrade-tsotsi reveals, the tsotsi as a figure was never disconnected from the struggle for liberation. Subsequently, he cannot be disconnected from any understanding of what a post-liberation society might entail.

Performing Thug Authenticity: The Contours of Tsotsi Embodiment

To discuss how Mandoza disidentifies with tsotsi performance, working within and between its signifiers, it is necessary to map the contours of this performance. First, the thug establishes spatial presence through marking himself as being from the rough areas of the township, in Mandoza's case being from Zola, Soweto. Zola is "widely regarded as the toughest neighborhood in Soweto. . . . There is a legend that if you were to run into a house in

Zola hoping for protection from criminals, you are likely to be given a panga [machete] to go outside and fight back. . . . [T]hat's the general view of Zola—where only the streetwise survive" (Tabane 2005).

The narrative of spatial authenticity is central to the making of the tsotsi performative, so it is unimaginable for the tsotsi to originate from nontownship space, thus marking the tsotsi as a particular racialized and spatialized performance. In the case of Mandoza, he established his thug spatiality through a variety of mechanisms. His first album, *9 II 5 Zola South*, reminds his audience that his origin is from the rough township of Zola. Furthermore, his township of origin was mentioned in almost every interview or press release regarding Mandoza's life experience. That this spatiality is inevitably connected with poverty and struggle is an important part of establishing the credibility of the tsotsi performance. In Mandoza's own words, "Well, Zola is the roughest part of Soweto. It was difficult to grow up under that situation and with those influences around us. Thuggish. It was all of that like, it was so difficult for us to grow up under that. We used to feel in love with being a thug, we all wished to be thugs when we grew up. Thugs were the only role models in our township you know" (Kaganof 2006b).

Mandoza provides a link between the spatiality of place: in this instance, Zola township and thug life. The narrative is also buttressed by the notion of criminality and redemption. As an audience, we are expected to sympathize with the already-present narrative of the Black township man who overcomes adversity to make something of himself. The narrative of overcoming only makes sense when produced through the spatiality, which marks its subject as synonymous with township poverty. The origin story and its production through space are critical to understanding the labor of tsotsi performance.

Street credibility via the announcement of a criminal past is also central to the performance. Importantly, the criminality (as seen above) is closely connected to the space of origin for the tsotsi. The possibility of danger and violence recounted through criminal behavior is key. Crucially, the nature of this criminal behavior typically limits itself to property crimes. While the performance of the tsotsi must suggest criminality and danger, it cannot be "too real." In other words, the tsotsi must be dangerous and capable of inflicting violence, but he should not actually be a reformed murderer or rapist. In the case of Mandoza, the criminality of his performance is heightened by the fact that he apparently served a prison sentence at the age of sixteen in Johannesburg's (in)famous Sun City prison. Mandoza gave the following account of his shift into a life of crime at a young age: "I was just a teenager when it happened, just 16. And you know, it was hard to resist the older guys in Zola, who were gang members and who were always pressuring the younger kids into taking part in crime. So I landed up in Sun City (Diepkloof jail) with the old timers for one-and-a-half years. It was hard" (Music.org.za n.d.). The township setting is reduced to criminality in these interviews, leading to a natural link between young Black township men and lives of crime. However, the primary purpose of these performative utterances is to establish the performer as a credible participant and witness of the roughest of township life. The result is that the performance persona takes on an aura of unquestioned authenticity.

In a separate interview, as he recounts his troubled teenage years, Mandoza reiterated that he was involved in the "thug lifestyle" at a young age:

> You are not a man if you are not involved in that direction [criminality or the thug lifestyle as presented by the interviewer]. And people are going to look at you like you're scared like you are a coward. You know what I am saying, stuff like that. We had to. We had to do that. I got into car theft. We stole cars and stuff like that. At that stage I got arrested. I sat for two years in jail. I sat in Sun City in 1992, the biggest jail in Gauteng. It was like that. But my heart was not there. I did what I did just for fun, just to get a name in the ghetto. But that wasn't my life. (Mandoza, in Kaganof 2006b)

For Mandoza, a life of crime is also linked to establishing heteromasculinity in township space. Mandoza presents his involvement in car theft as a rite of passage crucial to constructing notions of appropriate masculinity. He hints at the nonremunerative aspect of securing a reputation in the township by suggesting that his involvement in criminality was necessary to bolster his rep. Yet the representational and the material are linked and bolstering his rep in the township would have been important to building the street credibility essential to financial success within criminal syndicates. If nothing else, his reputation both for criminality and for surviving the notorious Sun City prison continues to bolster his musical career, even in its afterlife.

As conventional methods of securing masculinity through work fade for young township-based Black men, criminality becomes an increasingly attractive option for securing financial resources. As a source of financial stability, criminality for Black township-based men is nothing new. However, the turn toward neoliberal economics has meant that much of the growth that South Africa has experienced recently has occurred without significant increases in waged employment. According to one academic, "South African industry tends to be capital-intensive, investing in machinery rather than labor. . . . Many of the young people are not sufficiently skilled to fare well in the job market . . .

and the [capital-intensive] economy has no use for such people" (Kharsany 2013). Glaser also suggests that the experiences of the contemporary labor market in post-apartheid South Africa are gendered precisely because the expectations for securing masculinity through employment remain, even as the economy is not able to provide the kinds of manufacturing jobs that might have allowed young men to meet these expectations. "Many young Black men are facing a crisis of self-worth and identity. They cannot provide for families in circumstances where men are expected to be the providers" (Glaser, in Kharsany 2013). That Mandoza would shift his labor from the realm of the criminal to the cultural labor of kwaito is instructive for thinking about the avenues of work available for young Black township men and the central figure of the tsotsi as a performative identity that links crime and kwaito.

There is also an additional factor at work in the link between performative Black masculinity, criminality, and space. While recounting his prison experience, Mandoza also discursively reveals the machinations of carceral spaces that Black men exist in and work through. In this way, prison becomes a rite of passage for the tsotsi that bolsters his credibility. However, the narrative also simultaneously makes the carceral state seem inevitable, natural, and legitimate. Rashad Shabazz (2009) discusses how state and nonstate actors in South Africa and the United States apply the disciplinary and surveillance tactics of the prison to Black communities. In essence, Black people are primed for imprisonment long before they actually enter a prison. As Loïc Wacquant (2001) notes, these tactics of surveillance are gendered, linking Black women and children to the disciplinary structures of the welfare state while pushing Black men to criminality and the prison. These prison masculinities, as Shabazz (2009) has observed, influence the performance of Black masculinity outside of prison space, revealing a symbiotic relationship between Black masculine performance and carceral space.

In South Africa, this pattern can be observed through a division of the Black body politic into good subjects who are passive recipients of state beneficence on the one hand and unruly criminal bodies that discourage capital investment and disrupt capital accumulation on the other. Service delivery protests and the Marikana miners' strike are two examples of the media figuration of protest as Black (criminal) masculine embodiment. The perceived gendered nature of these challenges to the state allows them to be reconfigured as criminal and the participants as tsotsis, revealing the linkage between the Black men, gendered masculine performance, space, and criminality and punishment. The tsotsi performance as enacted by Mandoza relies on these linkages in ambivalent ways that simultaneously support and disrupt the equivalence of township Black men's bodies to criminality and imprisonment. For instance, while Mandoza trafficked in his time in prison to authenticate his gangster masculinity, he also suggested in the aforementioned interview that being a criminal is not really who he was for "my heart was not there . . . that [criminality] wasn't my life" (Mandoza, in Kaganof 2006b). In this way, Mandoza reveals the disidentification central to how he navigated the creation of his thug performance.

Mandoza established the performance of the tsotsi through vocalization and his use of language. The voice of Mandoza, gritty and gruff, establishes a dangerous street-based masculinity. Bhekizizwe Peterson (2003) notes how the hustling figure is often linked with that of the "dog," the symbol of hustling masculinity. Mandoza's voice then links to this theme and helps to establish his rough masculinity. In addition, language plays an important role in understanding Mandoza as located spatially in the township and located economically on the lower rungs. For example, Mandoza used language to establish that while his fourth album, *Tornado*, may have deviated sonically from the township in its production, it remained solidly *ekasi* in its lyricism. In 2002, Mandoza was embroiled in a controversy over his appearance on *People of the South*. As the interview proceeded, it was revealed that Mandoza was less than proficient in English and was stumbling through the interview badly. His appearance on the show became fodder for Black comedians and talk show hosts (most coming from the Black middle class) who made fun of Mandoza's imperfect English and used his lack of English fluency to mock him. Mandoza also alluded to his struggles with English in his role as a Black rugby player in the 2006 film *Number 10*.[1] However, the notion of not being able to speak English well serves as a marker of simultaneous marginalization and authentication. The lack of skills in English identifies the subject as authentically township and resistant to the narrative of middle-class, upward mobility. The fact that Mandoza succeeded despite his limited educational background and nonproficiency in English reveals him as a subject whom people who are in similar positionalities can relate to.

Through language, the tsotsi becomes the embodiment of the locally savvy figure who negotiates the latest trends within Afrodiasporic Space and contextualizes them for local realities. Perhaps this was the reason why the tsotsi figure was to become so representative of urban life, and thus so lucrative for those who perform it. In her article on quotable gestures, Heather Brookes discusses the importance of being kleva: "Being clever [sic] is of fundamental importance in every day township life. Admiring comments about who is clever and disparaging comments about who is not, using gesture and/or speech occur frequently in township talk, particularly among young men. Popular cultural forms such as kwaito, the music of township youth, make frequent reference in speech and gesture to being clever in songs about township life" (2001, 177).

The notion of being kleva is connected intimately with the idea of seeing, a type of seeing that marks the urban dweller as streetwise and therefore able to negotiate the dangers and pleasures of modern, urban life. This is contrasted with the notion of the *bare*, the fool, the individual who is not "with it." According to Brookes, the bare is marked as "someone who is boring because he cannot hold the attention of his peers on the street corner with skillful use of gesture and the latest iscamtho" (2001, 177). The tsotsi and his associated language then becomes the index of cool. Therefore, the growing acceptance and proliferation of what were considered criminal argots is not about an acceptance of criminality, but rather an admiration for the realities of socioeconomic struggle that almost all township residents are familiar with. The language of S'camto is contrasted with speaking what are termed "deep" versions of African languages. Speaking these languages in their standard form is often linked to an ethnolinguistic affiliation that is seen as inhibiting the ability of the speaker to integrate into urban, modern life. S'camto is also connected with movement and mobility as well as seeing. Urban Black subjectivity is understood through this notion of seeing that is abetted by "the ability to see in new ways as opposed to 'not seeing' or 'closedness'" (2001, 179). Ultimately, the tsotsi is a mobile figure whose adept use of language signifies both a literal and figurative command of the world around him and an ability to navigate through urban space even while being rooted in the township. Central then, to Mandoza's performance of the thug, are a number of quotable gestures, including "nkalakatha" and "skelegeqe," which mark him as simultaneously of the township but not confined by it. Recalling my earlier discussion of urban spatiality, the tsotsi, through the use of quotable gesture, can conjure up both the township and the spaces beyond the township. In this way, Mandoza's performances existed in and worked through an Afrodiasporic spatial imaginary.

Mandoza's performance was also marked by his perceived accessibility to his working-class, township-based fans. In a 2002 online chat, Mandoza was asked by a fan if he still ate *ikota*. An ikota is generally a quarter loaf of bread sliced open and then stuffed with French fries, pickled mango (*achtar*), cheese, and a choice of processed meat, such as bologna or hot dogs. This affordable sandwich is generally sold in township and inner-city corner stores and is considered Black township food. In his insistence that he still ate ikota regularly, Mandoza was affirming his connection to the everyday people of the township. The connection between food and identity in Afrodiasporic Space is central as it is commonly used to substantiate whether an individual has remained grounded. Commenting on a DVD highlighting Mandoza's performance in Sun City, fellow kwaito star Zola took the opportunity to emphasize how, despite his stardom, Mandoza remained grounded.

> There was never a time when Mandoza was too big for us. He always came home, talked to us, talked to the boys, he told us, "Listen I can't get you a recording deal directly but keep on pushing it's going to be fine," and when we went to shows and Chiskop was there and we were amateur artists we knew guaranteed that we were going to get on stage and perform. That's what this boy has done for us. He's an icon. Though he's my age I respect him. (Zola, in *Mandoza: Live in Concert*, 2004)

Zola considers Mandoza a role model because he maintained his ties to the township in general and to young township men in particular. This coded masculine space of mentorship positions Mandoza, despite his young age, as a father figure for other township men. Further evidence of his accessibility was provided through Mandoza's frequent trips to Soweto to attend community functions, as well as the sight of his vehicle moving through township space. I personally can attest to the stir that Mandoza caused when his car was seen cruising through township neighborhoods or when he showed up at a community event.

Accessibility is central to the brand of Black masculine performance enacted by Mandoza because the perception of accessibility buttresses the other elements of the performance, in particular the situating of the Black masculine body in township space and thus the authenticity of the tsotsi performance. We know that Mandoza is "really" of the township, because despite his wealth and fame he returns there regularly. Accounts of Mandoza sightings in township space serve to support the idea of familiarity and comfort with the township. That we were constantly reminded of this familiarity through numerous performative acts, including interviews, press releases, and song lyrics, suggests their iterative nature. Acts of accessibility must be constantly reproduced in order to counter the disjuncture between tsotsi performance on the one hand and musical stardom that signifies wealth on the other. Mandoza had to constantly balance the apparent contradiction between the two positions. In this respect, Mandoza is positioned similarly to other Black "thug" performances within Afrodiasporic Space. The moniker of "keeping it real" is often deployed in hip-hop parlance because it helps to negotiate the apparent contradictions in Black masculine performance. Accessibility is also integral to the positioning of Mandoza as an "example" and cautionary tale for other Black township men. He served as an exemplary subject because of his relatability, produced through what I am calling acts of accessibility. As we shall see, authenticity through accessibility will be a key component in revealing the redemptive possibilities forged through Mandoza's disidentificatory tsotsi performance.

A final contour of Mandoza's thug performance connects tsotsi performance to a sexually desirable body. To be clear, not all tsotsis are sexy, but as figures of Black masculine performance they are always sexualized. While a member of Chiskop, Mandoza was always physically fit. However, the change in his body from early Chiskop CDs to *Nkalakatha* demonstrates that Mandoza bulked up. His muscled body then became a major part of his performance, standing in contrast to other kwaito men performers, who often did not emphasize their sex appeal. His body was a way to secure heteromasculinity and instantiate his desire to women. This is similar to other thug performances throughout Afrodiasporic Space, such as that by DMX, whose popularity relied as much on his raw sex appeal as his street credibility. In fact the two are linked, since visual evidence of heteronormative masculinity secures claims to authentic thugness. The appearance of the half-dressed body can also lead to possibilities of homoeroticism, opening up the tsotsi performance to queer readings. During the height of his popularity, Mandoza regularly appeared shirtless in his music videos (beginning with "Nkalakatha") and in many concert performances. The display of the muscled, toned, and generally sweaty (in the space of performance) Black male physique connotes power, virility, and heterosexual masculinity. That we were allowed to admire and appreciate the Black masculine form that Mandoza himself was obviously proud of produces another part of his appeal. For while Mandoza was a thug, he was also a handsome, athletically built, and well-dressed thug (see figures 5.3 and 5.4).

As an example of the centrality of his muscularity to his thug performance, consider the media attention that was paid to Mandoza when in 2011 he appeared in public looking shockingly thin and with facial discoloration. In retrospect, we should understand this shift in appearance as potentially the early symptoms of the cancer that eventually took his life, but at the time the secrecy surrounding this shift allowed innuendo and rumor to flourish. The following comments from fans on the celebrity culture website justcurious.co.za provide samples of the kind of concern expressed about Mandoza's health:

> Mandoza babe, time to do something about his health problem ... whatever it is! This is not a good look! You look emaciated—a far cry from that hott body you once had!! DO SOMETHING! Siyakucela bhuthi [we are begging you, brother]. (purpleheart, in Brown Shuga 2011)

FIGURE 5.3. Mandoza advertisement, in which he is depicted with a smooth tsotsi masculine style. Source: Facebook: Mandoza Nkalakatha.

FIGURE 5.4. Mandoza depicted with a street tsotsi masculine style. Still image from the "Indoda" music video (2009). Source: Mann Made Media.

> A true moment of silence for Mandoza . . . a man dat was our own getto loving fantasy. (IBAR, in Brown Shuga 2011)

> I hope that he is well tho & that the emaciated state of body is not a reflection of any health problems. (Ms. Zie, in Brown Shuga 2011)

Three themes emerge from the commentary surrounding Mandoza's "emaciated" appearance that reinforce the importance of the muscular, toned body as central to his tsotsi performance. First, the commentary is centered on women's opinions about Mandoza's appearance. This is not to suggest that men did not express concern about Mandoza's shift in appearance; rather the spotlight focused on how Mandoza had become unattractive to presumably heterosexual women. Second, his new body appearance was contrasted with memories of his previous "hot" and "sexy" body. The 2011 body then is used to recall the previous muscular one and its sexual desirability as demonstrated in the following commentary:

> Looking at Mandoza, I got reminded of when I first moved to Jo'burg about 10 years ago. My then boyfriend, his friends, their girlfriends and I went to Calabash (those were the days). Later on Mandoza arrived looking so hot and hit on my boyfriend's friend's girlfriend (excuse the mouthful), telling her that he was there in a limo and had just come back from the SAMAS [South African Music Awards] where he won a SAMA and we should leave with him and ditch our boyfriends. The nerve! It almost caused such a fight. Shame yazi! (Pokerface, in Brown Shuga 2011)

The previous body of Mandoza is recalled and also reminisced about in a pitying manner. The combination of his celebrity and his body was so desirable that Mandoza could easily tempt a girlfriend away from her boyfriend despite the boyfriend's presence. This leads to the final theme concerning Mandoza's drastic appearance change: that the shift in his appearance, particularly the loss of his once-fit and muscular body, indicated health problems, specifically HIV. While there were those who defended Mandoza against such claims, suggesting that not all illness or weight loss is radically equivalent to HIV, more common were insinuations that Mandoza's obviously deteriorating health were symptomatic of HIV infection. The HIV body marks the constitutive outside of the desirable tsotsi "ghetto fantasy" that was central to Mandoza's performance. So potentially harmful is the association of HIV with the Black body that Mandoza felt compelled to publicly deny rumors of any infection, while maintaining an air of discretion over what, if any, health challenges might have been present. That HIV is incompatible with the public performance of the tsotsi is reinforced, however unintentionally, by Mandoza's denials. One reporter discussed Mandoza's need to go on national television to refute rumors of his HIV status, and that this served to reinforce the stigma around HIV: "Mandoza's rush to the television studio to dismiss the rumors was a panic move. It was the act of a cornered human being who understands that if he does not challenge the rumors, the whispering campaign will continue. As a person whose success relies heavily on his public image, Mandoza understood that his very livelihood was at stake in a society that stigmatizes those who are HIV-positive" (Mchunu 2008).

The disconnect between the public HIV-positive body as sick, diseased, and undesirable and the tsotsi body as healthy, sexy, and vibrant reinforced the complicated dynamic between Mandoza as a performer and Mandoza as a person. A particular kind of body was central to his performance and it did not include being emaciated, having facial blemishes, or being HIV-positive.

Rehabilitating the Tsotsi: The "Family Man" and "Focused Thug"

By analyzing the tsotsi as a redemptive figure, I consider the kind of labor that the tsotsi does in post-apartheid politics of redemption and reconciliation and for whom this labor is enacted. In other words, why does post-apartheid (Black) society *need* the tsotsi as a redemptive figure? What intraracial forms of reconciliation post-apartheid does Mandoza's thug performance enable? And how might young Black men like Mandoza who enter the arena of performance use this need for their performative labor, which admittedly limits and constrains their agency, for their own benefit? In order to analyze the possibilities of redemption, I assume the influences of commodification, the role of politics, and the importance of race, gender, sexuality, and class. Young, poor, and working-class Black men are expected to do and enact performances of redemption that are not required of global neoliberal capital or its allies among the South African political and economic elite. As Adam Haupt (2012) argues in his discussion of tsotsi imagery in post-apartheid media, "Major corporations did not have to answer for their crimes—with the notable exception being that the scale of their transgressions differs remarkably. Unlike Tsotsi, who finds redemption when he returns the stolen baby by the end of [the film] *Tsotsi* and, thereby, attempts to correct an injustice, corporations that benefited from apartheid have not been pushed to engage in any significant measure of introspection—they do not need to be redeemed" (181). Haupt aptly points out the hypocrisy in overburdening the body of the tsotsi with the theme of redemptive performance while remaining silent on the ways in which global capital does far more damage, historically and contemporarily, to South African society. However, the constraints imposed by neoliberalism and the contradictions global capitalism generates provide moments of possibility. It is these moments of possibility that are engaged through Mandoza's disidentificatory performances of the tsotsi.

This desire for the tsotsi's rehabilitation can be partially couched through the body of Mandoza, in this case the desire to recast him as a "family man," an identity that is intertwined with notions of class. The family man is the marker of urban Black bourgeois civility. Mandoza as a family man served two different projects. On the one hand, he represented the desire for the elite Black male subject (representative of the political ruling class in post-apartheid South Africa) to rehabilitate the tsotsi to the national project. On the other hand, he represented the fantasy of getting the tsotsi down the aisle for the (bourgeois) Black female subject. The ideal of the heterosexual marriage is a highly contested one among Black township women. In my observation, marriages are highly celebrated and the women who get married are envied and respected. However, there is growing evidence that marriage is increasingly seen as unattainable and therefore undesirable among significant numbers of township women. Because of the unemployment crisis, Black township-based men are less likely to be able to afford lobola payments and are less attractive as marriage partners. Furthermore, heterosexual township women insist on their autonomy and bristle at the idea that it might be lost within conventional heterosexual marriage situations (Ashforth 1999). Despite these realities, in my conversations with heterosexual Black women, marriage remains an idealized if not always attainable or realizable construct, as evidenced by comments on social media platforms describing Mandoza as a "ghetto loving fantasy."

Zola suggested that Mandoza serves as a model of responsibility and adulthood that should be emulated not only because of his mentorship of township youth but also because of his representation of the "family man." Commenting specifically about how Mandoza is an adult, and a self-made man, Zola states, "Him and I come from a situation whereby the family is broken at some point or another. Right, and he's mending it. Whatever went wrong with his family it's ending with him. And the day that I get married I will also end it. Because you have to be a family man at some time, he's grown" (Zola, in *Mandoza: Live in Concert*, 2004).

In this situation, Mandoza is considered the exemplary heteropatriarchal father. Zola relies on a discourse that posits children raised by single mothers or grandparents as products of broken families. Rather than see these nonnuclear family structures as offering support for communities, they are interpreted pathologically as lacking. In the United States, this discourse was popularly revealed in the Moynihan Report. The historical legacies of slavery, discrimination, increased urbanization, and (most controversially) Black matriarchy supposedly rendered Black families pathological (Moynihan 1965). In South Africa there has been a similar discourse that claims that colonialism and apartheid destroyed "the Black family." Part of the rehabilitation of Black masculinity in post-apartheid South Africa is the restoration of Black fathers to their rightful patriarchal places. The state, in recognizing same-sex marriage, has suggested that there should be space for alternative nonheteropatriarchal conceptualizations of family. Nevertheless, the power and continued attractiveness of the heteropatriarchal norm is demonstrated through Zola's commentary. Instead of analyzing forms of structural inequality that have widened in post-apartheid South Africa and therefore make marriage a far less attractive option for poor and working-class Black women, heterosexual marriage itself is offered as a panacea to the so-called ills of the Black family. Similar to a discourse in the United States that suggests that poor and working-class Black American women would benefit from heterosexual marriage, such an analysis fails to consider that women are remaining unmarried not because of some cultural pathology. Rather, they are making strategic choices based on the perceived unsuitability of such mates, due to forms of social inequality that include but are not limited to the collapse of industrial employment, the insecurity of labor in a postindustrial economy, and the prison-industrial complex. Part of what makes Mandoza eligible to be a "family man" is the fact that, unlike many other men from Zola township, he has economic resources that make him an attractive and suitable partner.

While the heteropatriarchal family construct has resonance among many Black men, women play an equally crucial role in this construction of masculinity. Thus the salience of the heteropatriarchal family construct, and the discourse of the township thug made good, is attractive to many Black women as well. *True Love* magazine, a women's lifestyle magazine that appeals to aspirant Black women, featured Mandoza on the June 2005 cover. Mandoza, dressed in a nice suit and tie, appears with the words "I am a family man" emblazoned on the cover. We are told that, inside, the magazine will feature exclusive pictures of Mandoza and his family along with an article about the tsotsi who has traded in the rough life for middle-class bourgeois domesticity. In fact, the lead of the article states, "Kwaito star Mandoza has long made headlines, as much for his wild past and former drug addiction as for his music. Yet since the birth of his two sons, he's gladly traded clubs and cocaine for nappies [diapers] and nursery" (Ramackers 2005, 74).

The article features Mandoza in a series of casual poses with his wife and their two children. The pictures confirm the dream of middle-class domesticity, with the nuclear family (not the extended family) as its centerpiece. His wife, Mpho, holds their youngest child in each of the pictures, while the oldest is posed below his father. The article's visual text supports the remaking of the tsotsi into respectability. Mandoza talked about the joys of being a father: "I wish someone had told me before about the wonderful experience fatherhood is, and the joy it brings. I can still hang out in the streets, chill with my friends and talk the language. But when I get home, I can just be a parent—I don't need to front it" (Mandoza, in Ramackers 2005, 74).

Mandoza acknowledges two different kinds of performances. First, he asserts the performance of Black masculinity in the streets and suggests that his performance is a front. In this case, describing the performance as a "front" acknowledges not the lack of verisimilitude to the performance but its constructed nature. Such everyday performances are a necessary part of negotiating the dangers of township life. Judith Butler's (1993) assertion that, within gendered performance, the instability of the form is marked by its need to be constantly reiterated is instructive here. This instability for Butler provides the very possibilities of destabilizing the hegemonic codes of gender. On the other hand, critics of Butler have argued that such reiterations, particularly for Black people, are not always hegemonic and often are critical to survival (Johnson 2003; Namaste 2000). People get killed and face danger, not because they fail to perform the tsotsi adequately, but because they are young, Black, working-class men in a racist, capitalist society. To paraphrase Butler (1993), their bodies are not those that matter. Consequently, Mandoza suggests that his performances on the street are a front, yet he also alludes to the fact that they are necessary. These performances were necessary whether for actual life or death, or to maintain a sense of street credibility central to his musical persona, and thus his livelihood.

In creating the home space as a place of sanctuary, where the performance of the tsotsi fades away, Mandoza also reinforced the performance of the family man. Heterosexual women are encouraged to imagine that the tsotsi can be rehabilitated away from dangerous Black masculinity and morphed into a protective, kind patriarch. For many working-class Black women, their male partners are not able to fulfill this role, and the state currently plays the benevolent patriarch role—even though very minimally—by providing child-support grants. Mandoza's wife, Mpho, is constructed as the ideal mother and wife, patient with Mandoza through his drug problems and numerous crises within their marriage. Her role in the profile is central to instantiating Mandoza as a rehabilitated subject:

> Mpho also looks back on her husband's addiction as a crisis in her marriage. "Initially, I tried to convince him to get help for his [drug] problem, but I realized that it was up to him to take the first step." . . . "Our relationship wasn't good at the time and it was very strenuous on the baby [Tokollo]," continues Mpho. "Many times I felt like leaving, but the talks I had with my father and other close family kept me going. People also reinforced my faith telling me there was no problem greater than God. And, of course, my love for Mandoza was the ultimate reason to stay." . . .
>
> "I'm very secure in our relationship because we've been through hard times before and when we got married, we knew we were committing to a lifetime together. Mandoza's a very different guy on stage to the man he is at home and I'm the one who knows him best. The real Mandoza—the one who leaves his clothes lying around—is the one I married, and I love him unconditionally," she [Mpho] says. (Mpho Tshabalala, in Ramackers 2005, 74, 76)

What emerges from this series of quotations is a dynamic through which Black women are positioned as managers of state and corporate anxieties about poor and working-class Black men and their place in post-apartheid society. Black women are implicitly assured that if they properly manage Black men for state and corporate interests, staying faithful and persevering through difficult relationships, the result will be the happy heteronormative ideal. Also as a housewife, she has a luxury that few Black women in South Africa can afford. Significantly the choice made by Mpho (to be a stay-at-home mother) is a nearly impossible choice for most Black women, even those of the middle class. Therefore, the "ideal" family—patriarchal father breadwinner, feminine domestic mother—is inherently constructed in ways that pathologize most Black familial structures.

My analysis suggests that tsotsi performance is not just a space of passive subjectification but rather an opportunity for forms of self-fashioning. Mandoza has taken this figure of the tsotsi and actively worked at creating something different out of it. The tsotsi is ultimately supposed to be an uncouth and irredeemable subject, who must be disciplined by the state policing apparatus. What Mandoza has done is to disidentify with this notion of the thug. When speaking about his song "Skelegeqe," Mandoza stated that a "skelegeqe" is a thug, but in the song, he says,

> Mina, I'm a top skelegeqe,
> Mina skelegeqe focused.
> Myself, I am a top thug,
> I'm a focused thug.

In the same interview he stated that while the song speaks of a thug, it speaks of one who is focused on bettering his life. Therefore, the suggestion is that being a thug is not completely irredeemable nor is it a static state of being; instead thug materiality can be changed. The focused thug, one who sees his life of criminality as a process rather than an end, reimagines ideologies about criminality, Black men, and township spaces. This is one critical way in which Mandoza remasters the iconography of the thug. Through mastery of the thug performance, he presents a slightly shifted copy. Yes, he is a thug but in being a thug, he suggests the possibility of so much more. The ultimate goal is to move from being on the side of criminality and marginality to being in a position of control in his life.

Mandoza offers that the unfocused tsotsi will eventually be consumed by a life of substance abuse and violence. In contrast, a focused thug can have different options in his life. These different options pull toward an identification with middle-class bourgeois respectability that I critique as limiting, which troubles for a moment the reading of the tsotsi as a disruptive subject. However, what made Mandoza such an intriguing figure is precisely those moments of excess that do not allow him to be the family man. For instance, there was an element of Mandoza's performance that always understands that, as a tsotsi, even one who is rehabilitated, he does not belong. This was demonstrated in the furor over what role, if any, Mandoza played in the death of R&B star TK, who, it should be noted, was from an elite royal family. While it was subsequently revealed that Mandoza did nothing to cause TK's death, it is also the case that while others were with her during the night before she mysteriously died, people were more suspicious about Mandoza's involvement. Despite being a family man, the fact that he was a young man from the township with a criminal past made for media frenzy and speculation. Mandoza felt that it was necessary to make efforts to clear his name. Therein lies the tension inherent in the tsotsi, both as a model of Black masculinity and as a subject to be tamed and put to the service of the post-apartheid nation. Mandoza shows how the thug is a human, changing his life into the narrative of upward mobility propagated by the post-apartheid state. On the other hand, because he is not well educated, does not speak English well, and is marked as being "township," he cannot ever truly access the middle-class respectability of the family man. The unassimilable aspect of the tsotsi makes performances of unrefined masculinity, particularly in Black communities, an attractive symbol. The unassimilable excess of Mandoza resonates with many Blacks, both middle class and working class and poor, either as a sign of authenticity or as a sign of the worrying prospect that many educated middle-class Black South Africans possess. At the end of the day, the Black middle class worries that there is always something excessively unassimilable about Blackness, and in celebrating Mandoza, perhaps that unassimilability, perhaps a slightly reformed "ghetto fabulousness," should be retained rather than hidden away or seen as a source of correction. Instead, those ideas and behaviors, which are supposed to be "corrected" with increasing economic capital, become a source of cultural capital.

I analyze Mandoza's 2004 video for his song "Indoda" in the context of the health, legal, and financial troubles he faced in the years before his death and his attempt to make a comeback via his 2013 album *Sgantsontso* and an e.tv reality show, *Rolling With*. In the video for the song, Mandoza appears to make himself into a literal consumer product. The formulation of Black bodies into literal consumer objects is nothing new. The history of chattel slavery and near-slavelike conditions for people of African descent in South Africa is well documented. Mandoza turning himself into a series of well-known South African brands, yet re-branding them with his moniker, might initially seem alarming, given the history of state and global capital control of Black bodies and Black productive labor. However, in the post-apartheid period, such bodily commodification can signify alternative readings.

Mandoza built a career trading on the image of the tsotsi. Yet through his performances, he has shifted this stereotype of Black masculinity from the margins of society to the center through his music and lyrics to the point where even wealthy white kids in Sandton are familiar with aspects of township lingo. Mandoza's performance of the thug represented a disidentificatory conscious manipulation and parodic reworking of the stereotype of dangerous Black masculinity for market consumption and his own enrichment. Indeed, turning the trademark tsotsi into trademarked South African goods and consumptive products brings the tsotsi literally into the homes of everyday South Africans, and it does so in a way that is transformative of the meaning of the thug. Shifted from irredeemable inhumanity, Mandoza's thug is humanized, wanting the same things as the rest of us and focused on turning away from a life of crime to obtain the good life. The song "Sqelegeqe" communicates the essence of a focused thug who reworked his life away from criminality, thus in many ways mirroring the artist's own transformation from lawlessness to bourgeois respectability.

In the "Indoda" video, Mandoza mapped his image and molded himself into a number of branded products familiar to most South Africans (see figures 5.5 and 5.6). At the beginning of the video, the camera focuses on Mandoza's face. He is inside a burning shack that, as he chants/raps, seems to fall away to reveal an informal settlement apparently engulfed by the same flames. His words forcefully remind us that a man can fall today, but tomorrow he will rise. As the fire burns around him, his face is imposed on a box of Lion Matches and a bottle of Black Label Beer, brands instantly recognizable to any township dweller. His face is placed onto a bank note that is then subsequently rolled up by Mandoza, alluding to his struggles with cocaine use. Later in the clip, his brand is imposed onto the newspaper classifieds where Mandoza is selling everything from old-fashioned record players to potent home cures promising strength and youthful vigor. While he continues to remind us to never give up, we see a boxing scene where Mandoza appears to fight the mirror image of himself. The video concludes with scenes of transportation and movement, with his brand appearing on a minibus taxi. In the next scene we see Mandoza himself riding in a train. Finally in the concluding scene, he drives his Chrysler Crossfire into the sunset.

Rather than being defined by the racial stereotype of the Black thug, this very stereotype is put into the marketplace where the highest prices are obtained for them (Botha 2006). "Mobility and consumption thus become the vehicles through which young Blacks control prevailing stereotypes and regain their individuality in the world" (Diawara 1998, 273). But what is Mandoza ultimately selling? Returning to the chorus of the song and juxtaposing it with images that include battles with himself, allusions to his troubled past, and symbolism of fire as destruction and rebirth, it would appear that Mandoza is ultimately selling the promise of redemption. Importantly this is not a redemption at the hands of the carceral state or toward the needs of the nationalist project. Instead, through the trials and tribulations of township life, Mandoza's thug redeems himself through his own guises. Central to the narrative is the understanding that forms of consumption and popular culture participation mark the movement of the tsotsi from the margins to forms of cultural citizenship that, as Nadine Dolby (2006) has discussed, create new forms of being that cannot be contained by the nation-state, revealing an Afrodiasporic affect. Equally clear through the narrative is the understanding that this is not a linear story, that successes can be followed by failures, and that one may have to fall in order to rise again.

I place Mandoza's 2013 struggles in tension with his tsotsi performance, particularly in relationship to the narrative of redemption and thus, following

FIGURE 5.5. A still image from the "Indoda" music video (2009).
Source: Mann Made Media.

FIGURE 5.6. A still image from the "Indoda" music video (2009).
Source: Mann Made Media.

the "Indoda" video, reinvention. While I do not mean to suggest that Mandoza's extensive troubles are somehow "performances," the way that he chose to engage the South African public regarding his misfortunes was performative. I have already alluded briefly to Mandoza's health problems and the way in which he was required to address them publicly. However, in November 2013, a report emerged in the *City Press* newspaper suggesting that Mandoza was "on the skids" and that he had lost his money, requiring a relocation from his posh suburban home back to Pimville, Soweto, the location of his mother's home (Malatji 2013). The article suggests that after a 2008 accident in which two people were killed, Mandoza was involved in a number of protracted court cases. While avoiding jail time, he was ordered to pay thousands of rands in restitution, which began a downward financial spiral from which he ostensibly could not recover (Malatji 2013). While the article does not mention his album sales, it is also the case that in this same period (post-2008) his album sales had declined significantly. In the article, Mandoza is described by neighbors as "friendly" and "accessible." Given that this particular media report emerged at the same time as Mandoza's new album release and an e.tv reality show chronicling his comeback, it might be useful to read his return to the township (note that the location of his mother's home is the decidedly more middle-class Pimville instead of Zola) not simply as evidence but also as performance.

In the reading I perform here, neither Mandoza's true financial situation nor whether he had really relocated to Pimville is of importance. Instead, what interests me is the role of the media and his management in constructing the story of his comeback and the necessary markers—financial hardship, return to the township, amiable relationships with regular township folk—that mark the appearance of this story and make it another layer to his tsotsi performance. Consider the following information from the article (Malatji 2013):

> News that Mandoza has relocated to the township was revealed by two music industry executives who are privy to the muso's private life.
>
> Another industry executive confirmed that the "Sgelegeqe" singer had moved but was too ashamed to tell them where he was living.
>
> Mandoza's wife, Mpho, confirmed that they had sold their Eagle Canyon house and that the family had moved to Pimville. But she declined to comment on the reason, saying it was "too personal."
>
> Neighbors, who spoke to *City Press* on condition of anonymity, confirmed that their "prodigal son" had returned to the township after falling on hard times.

Mandoza's manager, Curwyn Eaton, said: "Mandoza and his wife don't want people to know where they live. I know people have been asking questions about where they live but they would (be the ones to) divulge that. I'm also not going to comment on that because it is a personal matter, and I can't let you speak to him, speak to his wife."

Notice the main themes of accessibility, spatiality, and authenticity re-emerging in this account that is provided almost entirely by unnamed sources. As the reader we are encouraged to identify with the familiar narrative established by "Indoda." This is a nonlinear notion of redemption, success, and self-actualization. The tsotsi, reconfigured as the "prodigal son," returns home to reestablish himself and reconnect with his roots. This reconnection re-places him into township space, and through the testimony of neighbors we are re-assured of Mandoza's accessibility. Mpho once again labors on behalf of her husband's brand, for she is the only person who will go on record confirming the move to Pimville. Ultimately, the news report does the work of bringing the kwaito superstar back to the people, which will serve as the basis of his hopeful phoenixlike rise.

This rise was enacted through Mandoza's appearance on the show *Rolling With*. The show has served as a platform for a number of celebrities and public figures to promote their projects and redeem themselves in the public eye. Season 1 featured the kwaito star-actor Zola. Season 2 featured the singer/actress Kelly Khumalo. In both cases, the stars of *Rolling With* (Zola and Khumalo) typically had new albums to promote and public personas to rehabilitate. Mandoza was no exception, and the timing of his version of *Rolling With*, which premiered in September 2013, coincided with the release of his last album, *Sgantsontso*, in October 2013. Like all reality series that are not competition based, *Rolling With* purports to show the audience a slice of the reality star's life. In the words of the promotion for the show, "It's time for him [Mandoza] to tell his story" (*Rolling With*: Mandoza [season 1, episode 1]). Throughout the series, we are shown clips of a physically reinvigorated Mandoza performing for adoring audiences, recording tracks for his new album, hanging out with friends and family, and engaging in a number of practices (including driving a car in an obstacle course, receiving a spa treatment, and purchasing live sheep in the "ghetto") that work to instantiate his celebrity brand of redeemable thug Black masculinity. The sessions with friends and the purchase of livestock reveal him to be an everyday guy. The scenes with his wife cement him as a family man. The obstacle-course driving works to show Mandoza taking responsibility for his numerous transgressions involving car accidents. And the scene in the spa instantiates his denial of a "metrosexual" aesthetic in favor of what he terms the "gangster-sexual," reminding the audience of the realness of his thug performance and its sexual desirability. In a separate radio interview recorded for the show, the host remarks that no one would dare steal Mandoza's SAMAS [in reference to the recent theft of fellow artist Lira's SAMAS], and Mandoza replies, "You don't mess with me" (*Rolling With: Mandoza* [season 1, episode 7]).

Writing about the performance of gender in the ballroom community in urban Detroit, Marlon M. Bailey has the following to say about "realness": "to be 'real' is to minimize or eliminate any sign of deviation from gender and sexual norms that are dominant in heteronormative society" (2013, 58). "'Butch Realness [must be] unstoppable and unclockable [unrecognizable]. You look like a thug from the hood.' On other flyers, the criteria for the category reads 'Your mother can't tell, your father can't tell.' This means that the Butch must be 'unclockable' as a transgender man" (60). Bailey's analysis of realness performance among gender and sexual marginals in the urban United States can be extended to Mandoza's thug performance in South Africa by thinking of the ways that Mandoza used his public appearances to convince people that he is in fact capable of gangsterlike behavior. Thug realness is a form of performance that is racial, sexual, spatial, and thoroughly gendered. Mandoza shows the audience that the process of musical stardom has not changed the essential township-thug subjectivity. Hence, despite the need to portray Mandoza in ways that remove him from the idea of the thug as an irredeemable figure, the performance rests on the threat embodied by the tsotsi. In this case Mandoza's "thug realness" is instantiated through speech acts re-performed on his reality show. The realness of his performance provides the framework for understanding both his legibility and eligibility as the redeemed thug (see figure 5.7).

Remastering Masculinity: Queering the Tsotsi

"He's a bit of a tsotsi," my friend gushed. "But that's what makes him so hot. Doesn't he look a little like Mandoza?" The power of tsotsi performance lies in its ability to mobilize multiple desires that fulfill various social needs in post-apartheid South Africa. Mandoza's erotic performance of the tsotsi created new visibilities for other young men, whose performances of masculinity were similar to those of the kwaito star. Specifically, the tsotsi performance that Mandoza trafficked in created new pleasures, which allowed other tsotsis to emerge in post-apartheid society as objects of desire. Previously confined by location and class, the newfound dissemination of tsotsi performance through numerous South African media (including most importantly kwaito) facili-

FIGURE 5.7. Mandoza portraying the "focused thug" tsotsi persona (2004). Source: The CPP Record Company.

tated additional spaces of encounter for tsotsi desirability. These encounters enacted intraclass contact that pushed beyond the normative understanding of tsotsi desirability as a form of illicit eroticism (Miller-Young 2014). Instead the tsotsi masculinity could now be performed and admired by those who were not necessarily of the tsotsi class and location status as well as those for whom tsotsi performance engendered queer desire.

I joined a group of friends for a night of booze and fun at a shebeen that was down the street from the more middle-class Rock. In fact many of these friends might variously attend the Rock and also this nameless shebeen. The shebeen was described to me thusly: "It's a rough ghetto place, lots of rough guys are there, but a lot of them are down to hook up with us gays. In fact we come here to get rough ghetto boys." It seemed that this particular shebeen was preferred by my friends for two reasons: the accessibility of low-priced drinks, and the accessibility of hypermasculine gender–performing men who might be available as sexual partners. Queer men based in Soweto have written about their frustration with the limiting forms of masculinity available in township spaces (see Molobye n.d.). The discussion that follows is not intended to negate the real dangers that queers face from men who enact varying degrees of tsotsi masculinity. Nevertheless, during my numerous forays into township nightlife, I was to learn that tsotsi performance was a trait that many of the queer men I socialized with desired. What possibilities might have been available for the queer men I encountered within the confines of tsotsi masculine performance?

As we entered the shebeen, the mix was heavy on kwaito as the soundscape for the evening. In fact typical fare from the likes of new and old kwaito artists were played along with an occasional house song. As was usual of a township shebeen, we ordered a crate of dumpies (750-ml bottles of beer). These were placed in the center of our circle where empty crates also served as our seating. While there were a few women present, the ratio of men to women was easily 3:1. As I was told by many women, a rough shebeen, where drinks were cheap, security was lax, and the clientele was mainly poor and working class, would not be a place that an average woman would venture as the danger of these spaces was palpable. This did not stop women from entering these spaces entirely, but it did mean that the women who entered were often in groups or accompanied by boyfriends as many of the women I encountered that night were. Single women (in pairs, groups, or rarely on their own) could expect to draw a great deal of attention, some of it unwanted, and it was the inability of women in public space to control unwanted attention that gave the space its dangerous quality.

The combination of the gender-ratio imbalance and the class positionality of the men present, however, made many queer men feel that they were at a competitive advantage vis-à-vis more middle-class venues. Here, the ratio of men to women would be more in their favor and their limited disposable income could be harnessed to social advantage. While many of my friends were working class or lower middle class and lived in the township, they were also more likely to be engaged in formal steady employment, guaranteeing them a steady income that many township men did not possess. This often meant that in these less middle-class spaces, they would often command attention because they could afford to buy large amounts of food or drink at the shebeen in comparison to many of the other regulars. My friend explained, "The way to get these rough boys is to buy them drinks. You don't discuss sex or hooking up or anything. In fact before taking them home there may be little discussion or interaction between you two, perhaps a little flirting or small talk. . . . A lot of these guys come to the shebeen with little money for drinking, and in some cases with no money. Sometimes, their drinking buddies and/or friends will cover their tab, but at some point, that option is not available; that's when they come to us. They know we'll buy them drinks or allow them to share ours."

From what I understood, there was a tacit understanding that sexual activity could be a possible outcome of the exchange but was never guaranteed. Also, performing one's queerness too openly or flamboyantly could invite violence, given that the exchange occurring presumed a certain level of discretion. However, the dance floor was one place where explicit queerness could be performed. My friend said, "So we kind of use the dance floor to get the guys' attention. Obviously, for some of us due to the way we carry ourselves or the way we dress, it might be obvious that we are gay. But for others of us it is not so obvious. So the dance floor allows us to show that we are gay but also that we are available. It gives them something to look at, something to be enticed by." My friend explained that the dancing is often excessive, meant to be flamboyant and draw commentary and attention. As he finished explaining this to me, I glanced over at my friends who were already commanding the dance floor. As the kwaito house mix continued, they gyrated and gesticulated, bringing attention to themselves partially because of their kinetic energy, but also because the athletic dancing they engaged in had a decidedly feminine energy. This was for me the example of the nation dancing like Lebo Mathosa, although I imagine Makgoba did not have my queer friends in mind with their braided hair flying, their torsos twisted, their hips jutting out, when he suggested Mathosa's dancing as a nation-building strategy. I noticed immediately that the dancing had the effect of drawing the attention of the men in the shebeen, who commented, whispered, or pointed at the "gays" acting up. My friend continued: "It is fine that they point or talk, it means they are interested. Those will be the guys that will slyly approach you as you are going to buy another drink, smoke a cigarette, or go to the toilet. The group that is noticing you the most, is the same group of guys that will ask you for a drink. They are the same ones who will go home with you at the end of the night."

As I have discussed elsewhere (2014), these games of tacit misrecognition seem to me always tinged with the edge of danger. What could be enabling about this kind of masculine performance? As my friend explained, "Why wouldn't I want the same kind of thug? Why wouldn't I want a guy like Mandoza? I understand there is power there. I also understand that nobody would ever think that kind of guy could be gay or bi or whatever. Let's be honest: a lot of these rough township guys that act so masculine like to mess around. . . . They just don't want anyone to know about it, but a lot of them are bi, a lot of them experiment, hell, a few of them are gay! I like a masculine straight-acting guy and these township thugs give that to me." There is much to examine here, be it the preference for stereotypically masculine behavior, or the tensions of class that arise concerning who has the money to buy drinks and the assumptions of exchange that regulate the interaction. However, what I am interested in is the way that queer men identify the tsotsi as a desirable type of masculinity that they too can access. Ultimately the tsotsi performance of someone like Mandoza circulates in ways that mobilize queer desires and provides the space for vernacular tsotsi performance to access unconventional modes of desire.

As the evening wore on, the young man who had been pointed out by my friend sauntered to our section and began engaging in conversation with my friends. His S'camto was far too complex for me to follow along, but he seemed particularly interested in our group. At some point I caught a few words that he addressed to my friend: "I don't really know about this gay thing. But you guys seem nice, you seem cool. I like hanging with all of you." My friend was completely smitten, understanding that this admission was a sign that his overtures for the evening (his kinetic dancing, his femme attire, his offer of drinks) had done the requisite job of facilitating this encounter to its erotic ends. As the evening wore down, the conversation between the two grew more hushed, sly glances were exchanged, and the two left with each other for the evening. In this moment, multiple remastery occurred. For my friend, he remastered the space of the shebeen for his own erotic desires as well as the subjectivity of queerness, which might suggest that queerness either does not belong or should be met with violence in relation to tsotsi masculinity. For the young man who was the object of desire, his own desires might have been reimagined in this space; through that encounter, he also might reimagine the space of the shebeen as a site of nonnormative desires. The heteronormativity of thug performance thus becomes remastered, made queer, in this moment of possibility.

In his groundbreaking study of post-apartheid South African mediascapes, Haupt devotes attention to the amount of emphasis placed on Black masculinity "as an embodiment of the heterosexual, streetwise, gangster" (2012, 153). Haupt argues that contemporary market forces constrain the types of agency available for Black men performers. "The key issue is that the image of the Black man as thug/gangster/tsotsi has been commodified to such an extent that the agency of Black male subjects and artists to construct their own personal trajectories is somewhat constrained by market forces" (2012, 154). While I acknowledge that the focus on the tsotsi can have problematic consequences for a range of Black men and masculinities, and while I agree that contemporary South African mediascapes place an undue emphasis on the tsotsi as *the* form of embodied Black male performance, I depart from Haupt's analysis by seeking the productive possibilities within these limitations.

To understand my analysis of tsotsi performance in post-apartheid South Africa, I have relied on a reading of the kwaito artist Mandoza and the forms of tsotsi masculinities he mobilizes. Mandoza's performances are a form of cultural labor that must work within and through the commodified figure of the tsotsi. The limits and possibilities entailed within are part of the productive labor that Black men and Black masculinities are required to perform. Disidentification, the notion of a subjectivity that is neither totally oppositional nor totally capitulative, informs how young Black South African men like Mandoza perform their masculinity. It is within the struggle over how the tsotsi is performed, by whom, for whom, where, and to what ends that the productive possibilities of this admittedly limiting representation can be imagined. Mandoza built a profitable career navigating both the possibilities and limitations imposed by the tsotsi as an essential figure of the contemporary South African mediascape. He remastered the tsotsi, allowing the performance to circulate with a newfound cachet and larger circle of influence. This enabled the tsotsi, as a figure of kwaito performance, to engage forms of queer desire.[2] Whatever freedom may look like and however it may be performed, the tsotsi will be an important part of determining its meaning, and thus the practice of tsotsi performance is essential to understanding intersectional freedoms post-apartheid.

6

Mafikizolo and Youth Day Parties

(MELANCHOLIC) CONVIVIALITY
AND THE QUEERING OF UTOPIAN MEMORY

Kwaito started because we wanted something fresh for the youth. After Mandela came out in 1991 we wanted to make happy music, before those songs were about suffering and struggle.
—MANDLA MOFOKENG, from the band Trompies

In this final chapter I focus on the ways in which kwaito bodies engage in processes of historical memory and reconfiguration. These engagements not only challenge the notion promulgated by kwaito artists themselves that their music and performances are apolitical, but also illuminate how kwaito seeks to constitute itself with local cultural resources. Engaging with history affords young Black South Africans an important avenue to challenge critiques that their culture is simply about tuneful, entertaining performances with good beats. Instead, through their engagements with history—its events, traditions, practices, and tropes—kwaito bodies can articulate notions of freedom that emerge out of their reinterpretations of the past. The refashioning of collective and racialized memories enables young Black South Africans to engage in larger debates about freedom in the post-apartheid moment. Through remastery of apartheid memory, kwaito provides a space for the queering of utopian memory.

Drawing on a set of performances associated with the kwaito group Mafikizolo and the parties associated with Youth Day celebrations on June 16, different forms of kwaito performance appropriate and "remaster" memories of the past. I use two different moments in history—the 1955–63 destruction of Sophiatown, a mixed-race community located in Johannesburg that contravened apartheid's ideology of separateness; and the June 16, 1976, Soweto student uprising—to explore the relationship between kwaito bodies, the representation of memory, and the national present and future. In showing that members of Mafikizolo and kwaito fans are acutely aware of the past, these performances of memory demonstrate how the current generation seeks to free kwaito from the material, aesthetic, and social limitations imposed on it. At the same time, Afrodiasporic Space becomes readily available for cultural resources from which kwaito artists draw in their discourses and practices surrounding music, memory, and performance.

While Mafikizolo's reworking of memories of Sophiatown seems to have been received with nearly universal acclaim, the performances of the numerous June 16 parties demonstrate the precarity, fragility, and contested nature of memory. The numerous June 16 parties held every year are a source of condemnation, particularly from the Black political elite, who insist that June 16 should be commemorated with solemnity. The celebration of the student uprising poses particularly vexed questions about how pain and trauma are supposed to be performed. What does it mean to connect the suffering Black body of June 16 to the pleasurable kwaito body of June 16 parties? The parties do in fact commemorate the liberation movement through the exuberant celebration of life and freedom that are so prominently on display in these celebrations. These parties reveal June 16 to be not only about suffering but also about defiance, the power of youth, and the humanity of Black bodies. The performances central to June 16 parties consistently renew these ideals and provide a space for kwaito bodies to live out the possibilities fought for by those who rebelled. That this is misrecognized by the current political elite reveals that we ought to listen and take seriously the voices of the Y generation.

Throughout this chapter, I mobilize two concepts that are helpful in understanding the ways in which kwaito performance creates the space to engage in historical memory and refashioning: musical memory and melancholic conviviality. Musical memory builds from Ananya Kabir's (2004) concept of "musical recall," which she develops from Ron Eyerman and Andrew Jamison's (1998) concept of recall in relation to music and social movements. They write, "[Music] creates a mood—bringing it all back home, as Dylan once said—and in this way it can communicate a feeling of common purpose, even among actors who have no previous historical connections with one another. While such a sense may be fleeting and situational, it can be recorded and reproduced, and enter into memory, individual as well as collective, to such an extent that it can be recalled and remembered at other times and places" (161–62). In an article describing the function of contemporary Punjabi music in the UK Punjabi diaspora, Kabir (2004) notes how Punjabi popular music is able to express new Punjabi identities that recall a pre-Partition Punjab: such musical recall has a redemptive and commemorative potential inherent in its ability to bypass narratives of violence and nationalism and articulate instead post- and transnational modes of identity formation and cultural belonging.

Music provides a basis for social remembrance through notions of recall to signify critical engagements of the past that are political for the kinds of pasts they are able to mobilize in the present. Thus musical recall functions as a form of collective memory that is capable of invoking historical moments that preceded incidents of "cultural trauma." Defined by Eyerman (2001), who used the term in connection to transatlantic slavery, cultural trauma is "a dramatic loss of identity and meaning, a tear in the social fabric, affecting a group of people that has achieved some degree of cohesion" (2). Similarly, the recollection of the music and style of 1950s Sophiatown is predicated upon the moment before its destruction. Mafikizolo is one group that uses what David Coplan (2000) refers to as a "performative reconstruction of the past" in order to demonstrate the essentially embodied nature of aural popular histories. Through their performative reconstruction of Sophiatown, Mafikizolo enacts a form of musical memory.

Second, I rely on the notion of "melancholic conviviality" to explain the important role that the past plays in all aspects of contemporary South African public culture. Melancholic conviviality is an extension of Paul Gilroy's concept of conviviality, which "refer[s] to the processes of cohabitation and interaction that have made multiculture an ordinary feature of social life" (2004, xv). Rather than evoking spaces where difference is erased or harmonious, Gilroy's conviviality captures moments where rigid conceptualizations of race, class, gender, and sexuality become temporarily unstable. Gilroy suggests that for post-millennial Britain, conviviality is the opposite of melancholia built on racial exclusion and white supremacy. However, I couple the terms because much of the memory of South Africa's past is based on a nostalgic remembrance of a past that never was, with a particular emphasis on racial tolerance, class cooperation, and social order. In this sense, Black South African melancholia is not based on white supremacy but on a notion of interracial, intraracial, and interclass conviviality that evinces what Jacob Dlamini calls "native nostalgia" (2009). Contemporary invocation of a utopian vision of the past is a distinctive post-apartheid South African form of melancholic conviviality.

Representing Sophiatown

Much work documents the historical importance of Sophiatown in apartheid cultural lore. In this discussion, I highlight the importance of Sophiatown both materially and culturally for Black South Africans. Founded in the early 1900s, Sophiatown was initially meant to be an area that serviced low-income whites. However, its proximity to sewage treatment plants, distance from the city center, and poor drainage made the space unattractive for many white families, so plots of land were sold indiscriminately to whoever wanted them. Eventually, an urban, predominantly Black, yet racially mixed neighborhood formed in Sophiatown. Importantly, Sophiatown was exempted from ownership restriction laws of the Urban Areas Act and remained the only place within Johannesburg where Blacks could own property. As a result, while the majority of Sophiatown's residents were poor, Sophiatown had a significant property-owning Black middle class. "Ownership of real estate gave Sophiatown a sense of community, with institutions and a social identity that served as a defense against the dehumanization of the labor system," Coplan explains, noting that Sophiatown shared many of the social and economic conditions of newly formed townships specifically for Black residents but "offered a greater sense of permanence and self-direction" (1985, 144). Its proximity to the city of Johannesburg, in contrast to the newly built dormitory communities in Soweto, also made the site a particularly attractive space for Black South Africans.

Sophiatown has left two important legacies to the history and memory of South Africa (one material, the other cultural), and each functions as a powerful tool of remembrance. Etched in the imagination of the post-apartheid South Africa as a place where Blacks could own property, Sophiatown is mythologized as a space of traumatic origin and remembrance for those dispersed to the newly formed townships of Soweto, Meadowlands in particular, but also areas such as Dube and Orlando West. Because Sophiatown is often cited as a place where Blacks, Indians, Coloureds, and whites lived in relative harmony, its destruction, according to Rob Nixon, "obliterated the most diverse and culturally innovative community South Africa had known" (1994, 11). In Sophiatown, Blacks worked as doctors, owned businesses, and existed on relatively equal terms with members of other racial groups. "Sophiatown," Loren Kruger relates, "was an actual but thoroughly imaginary place that came to symbolize a fragile moment of racial tolerance and cultural diversity" during apartheid (1995, 60). In addition, Sophiatown is held up as a mythic space of class cooperation. Ulf Hannerz (1994), Coplan (1985), and Don Mattera (1987) all note that those who were well off lived side by side with those in rusty shacks, the communal water taps serving as a space of interclass cooperation and encounter.

However, it is important to understand that while Sophiatown was all these things representationally, realities were more nuanced. Scholars such as Coplan have noted that nearly 80 percent of the property in Sophiatown was owned by whites (1985, 1444). Furthermore, Mattera (1987) notes in his autobiography that the apartheid government stoked tensions among Coloureds, Indians, and Blacks in ways that belie the easy characterization of race relations in Sophiatown as harmonious and tolerant. Lastly, the fact that many lower-class and poor residents of Sophiatown in some cases eagerly relocated was connected to the sense of disempowerment they felt in the space. I detail these points not to dismiss the sentiment created by Sophiatown, but to emphasize the selective nature of its memory. What gets remembered and how can tell us much more about South Africans' contemporary needs and concerns than the reality of 1950s Sophiatown.

In this light, Sophiatown's cultural production is particularly noteworthy. The writing and cultural blossoming of the time period was known as the Sophiatown Renaissance, in direct reference to the Harlem Renaissance. The locally based jazz music that flourished in Sophiatown was initially dismissed for its lack of connections to rural, tribal identities being promoted by the apartheid government. The systematic use of Afrodiasporic culture both had historical antecedents and was a strategic maneuver to rework and re-create African identity in 1950s South Africa, as distinct from the tribal ahistorical visions created by the apartheid government. While the ability to play American jazz standards was highly prized during the 1950s, South African remastery of these standards evoked Afrodiasporic relationality. Hugh Masekela explains,

> Every African, African American that made it that we knew about, Duke Ellington, Louis Armstrong, was all through music. And, ah, these were Africans that were brought there, but their only way out of the plantation was through their musical prowess. You know. And there was no difference between like—to me there was no difference—between African Americans who went to the States and came out to the plantations, and Africans who were taken away from their lands, into the plantation which is where we live in South Africa right now. (Masekela, in Ballantine 1999, 3)

In comparing the colonial apartheid economy to a plantation, Masekela reveals how displacement and loss, experiences typically sutured geographically to the Americas, become a function of everyday life for Black South Africans. One of the few sites apartheid authorities could not fully control, artistic production became an important mechanism through which displacement and loss were contemplated and new futures and possibilities imagined. While jazz's transatlantic connections contributed to its initial lukewarm reception in South Africa, Sophiatown jazz became increasingly hybridized and Africanized. The music of Sophiatown (as well as writing about it in venues like *Drum* magazine) exposed Black South Africans' need to represent the specificities of Black South African urbanization and/as modernity, and it called on Afrodiasporic routes of remembrance (Holsey 2008) to battle the cultural essentialism of the apartheid state.

However, Sophiatown exists in a space of two different kinds of diasporic memories. The first reveals an Afrodiasporic flow of production that is neither one way nor teleological, where constant, continuous Afrodiasporic cultural exchanges are reworked on the African continent to be made relevant to local concerns and aspirations. Africa is therefore part of these very diasporic, dialogical, and polyphonic flows of culture and politics, rather than a static site of origin (Matory 2005). However, the trauma of removal—particularly the cruelty of forced removal and destruction that characterized Sophiatown—makes the community a further site of diasporic longing. Its residents dispersed to far-flung locations, the remembrance of Sophiatown is very much couched in pain, trauma, and physical dislocation similar to the "cultural trauma" described by Eyerman and Kabir. Consider this description as Mattera watched the destruction of his home in Sophiatown, which had been in his family for three generations:

> The vehicles were driven by two Boers who each had an African attendant sitting behind them as if they were apprentice bulldozers. . . . The machines began their destruction. My eyes were fixed on my grandparents' house. One of the killers attacked the kitchen, leaving a gaping wound in its side. Beaten and battered the kitchen collapsed and died. . . . I saw particles of blood and bone that were not visible to the naked eye; not anyway, to the dust filled eyes of my relatives. A strange incredible sadness came over me, like the sensation that comes to the skin when an insect crawls over it; or the sudden shudder at the sight of a decaying corpse. (1987, 17)

Mattera's personification of his destroyed house as a murdered person reveals that Sophiatown was not simply a collection of buildings but a community made up of a number of affective attachments. The notion not just of a building being destroyed but of a sense of communal and personal death pervades many of the narratives of Sophiatown. Bloke Modisane describes the destruction of Sophiatown in a similar fashion: "something in me died, with the dying of Sophiatown" (Modisane, in Kruger 1995, 64). Significantly, the displacement experienced by those relocated from Sophiatown was both a physical displacement, in the actual removal of their bodies from the space, and a psychic, spiritual, and emotional displacement as experienced by the gut-wrenching accounts of loss, the cruelty in the destructions of home and community, and the sadistic renaming of Sophiatown into Triomf, the Afrikaans word for triumph. Thus diaspora functions in relationship to Sophiatown on several vectors, through the use and manipulation of Afrodiasporic cultural forms, creating and existing in a space of what Paul Gilroy (1993) has termed "diasporic intimacy." However, diaspora was also in effect in the literal displacement of up to 3.5 million Black South Africans from their homes, and the emotional displacement of loss and terror created by the destruction of communities.[1]

In 2006 the African National Congress (ANC)–controlled government in the city of Johannesburg officially renamed Triomf back to Sophiatown. This was a culmination of an almost decade-long process that began in 1997 with a resolution to rename the suburb. The renaming ceremony included several dozen former residents of Sophiatown, including luminaries such as Adelaide Tambo and members of the Xuma family. The Xuma family home, currently a historical landmark and one of only three buildings to escape demolition, was honored with a heritage plaque.[2] Residents marched to Christ the King Anglican Church, made famous by Father Trevor Huddleston, and laid a wreath to commemorate him. The joyous celebrations, however, remained tinged with the traumatic memory of removal. Irene Kau, one of the participants in the ceremony and a former resident, stated, "The pain does not go away; every time you come here it is revived" (Kau, in Davie 2006). A council member of Ekhuruleni, Thabo Matsho, who was born in Sophiatown, expressed the frustration of walking through areas where his family once lived: "I felt like knocking on the door [of the house his parents once lived in] and saying, 'My parents lived here. It's my parents place. I want it back!' But you can't do that" (Matsho, in Chibba 2006).[3] Johannesburg executive mayor Amos Masondo spoke during the dedication ceremony about why Sophiatown remains etched in the memory of South Africans: "Sophiatown evokes memories of a vibrant, creative, multicultural community. A place where artists, writers, and musicians flourished, against all odds, in an atmosphere of racial tolerance. . . . [F]or the white establishment, the threat posed by Sophiatown was cultural as well as political. Sophiatown was a grand experiment in the management of cultural diversity" (Masondo, in Chibba 2006). For Masondo, as well as many commentators on Sophiatown, the memory of Sophiatown is not only an active episode of reconstruction, it is also a promise of a future that is racially diverse, tolerant, and prosperous: "Sophiatown is the past we dare not forget, [as well as] the future we must invest in" (Masondo, in Davie 2006).

Therefore, Sophiatown looms large in contemporary cultural politics because it has represented a nostalgic remembrance of a fantasy that many hope can become possible in post-apartheid South Africa. I call this nostalgic remembrance, or "melancholic conviviality." The melancholy of Sophiatown comes from a specific form of nostalgia produced within Black communities, which conceives the racial diversity of Sophiatown as a possibility of what could have transpired in urban South African space had apartheid not been instituted. It is a melancholy that simultaneously transposes itself onto the memory of Sophiatown, while speaking to the hopes and aspirations of the contemporary post-apartheid moment. Like Dlamini's concept of native nostalgia (2009), melancholic conviviality speaks to the desire for forms of certainty, social order, and predictable outcomes that seem absent in the post-apartheid space. This future-past reveals the anxiety about the contemporary moment. Given a second chance, will South Africa's diverse population achieve the hope of Sophiatown that was, simultaneously, never in existence and never fulfilled? This anxiety about the future of South Africa is intimately connected to the cultural politics of popular music cultures, particularly in relationship to young Black South Africans. Thus, the way that kwaito bodies manipulate the past through numerous citation practices is of importance as a gauge of memory (see figures 6.1 and 6.2).

Mafikizolo and the Performance of Musical Memory

To discuss musical memory in kwaito, I focus on one kwaito group, Mafikizolo, and how they put imagery of Sophiatown to use in four post-2000 albums: *Sibongile* (2002), *Kwela* (2003), *Van Toeka Af* (2004), and *Six Mabone* (2006). Their use of Sophiatown in the totality of their musical practices (dance, fashion, and sound) exhibits that, far from being disconnected, young kwaito stars are quite cognizant of history. Their citation practices reference moments such as Sophiatown, both as a site of critical remembrance and as a space for possibly configuring and reconfiguring the present and the future. Furthermore, these citation practices pay homage to previous generations of musical artists who often struggled to create music under far more trying circumstances than

FIGURE 6.1. Sophiatown residents' protest to displacement occurred through a continued habitation despite post-state destruction. Source: Jurgen Schadeberg (1955).

FIGURE 6.2. Residents occupying the rubble of Sophiatown.
Source: Jurgen Schadeberg (1955).

those that exist today. Therefore, the genre of Afro-pop, which Mafikizolo is exemplifying, is a critical and conscious engagement by kwaito artists with South African history (see Fourie 2005).

The desire to re-create a sense of Africanness through tradition connects Mafikizolo with the concept of melancholic conviviality that I describe and that forms the bedrock of musical memory. However, the tradition that Mafikizolo accesses, and the manner in which they perform it disturbs static conceptualizations of tradition. According to Nhlanhla Nciza (née Mafu), one of the lead singers of Mafikizolo, "Kwaito is so broad, when you look at kwaito you look at it as an umbrella, there's other music coming out from that umbrella of kwaito, like you find your Afro-pop, you find your afro-house, you find your traditional music under kwaito as well, so that you're not limited you can play around with music and the instruments and the lyrics as well" (Mafu, in Kaganof 2006a). Here Nciza reveals one of the most important aspects of kwaito and South African youth performance in the post-apartheid period: its dynamism and its ability to account for a dizzying array of influences. The sense that within kwaito performance limitations and boundaries can be pushed is essential to understanding the evocation of the past in contemporary youth performance and its possibility to be remastered and queered.

The three members of the original group, Theo Kgosinkwe, Tebogo Madingoane (now deceased), and Nhlanhla Nciza, joined together in the mid-1990s. Kgosinkwe and Nciza knew each other as neighbors growing up in the West Rand township of Kagiso. Madingoane, a veteran of the music industry who had worked as a dancer for famous 1980s artists such as Chicco, also joined the group. According to Nciza, the group initially attempted to record a number of tracks for a demo tape that featured predominantly R&B music. However, one track from the demo tape was a kwaito song, and apparently it was enough to impress kwaito-house super producer Oscar Mdlongwa. The group's shift from a focus on R&B to kwaito would prove indispensable for their initial success, given the fact that locally produced R&B has never been a big seller in the South African market.[4] However, the fact that the two main leads in the group sang was noteworthy given the landscape of kwaito in 1996, which emphasized primarily all-male groups and male solo artists who chanted their lyrics. The centering of Nciza's voice and image, as the lone woman of the group, was also an important departure from the kwaito landscape of the mid-1990s, as well as her far tamer, less sexualized presentation of femininity compared to the women of Boom Shaka and AbbaShante.

Their first album, a self-titled EP, featured the moderate hit "Mafikizolo." The group chose the name Mafikizolo, which can roughly be translated as "the new kids on the block," to express their youth and freshness. At this point, the group seemed to be following in the footsteps of successful mixed-gender groups such as Boom Shaka, with an emphasis on electronic slowed-down house beats as the driving force behind the female vocals. In 2000, the group came out with their second full-length album, *Gate Crashers* (see figure 6.3). This album featured the massive hit "Lotto," which was eventually remixed by house pioneers Masters at Work and distributed internationally. "Lotto" was their biggest hit on the dance floor. The song had sparse lyrics and emphasized the group's dancing and choreography skills, something that would serve them well as they developed their craft, but which nonetheless did not allow Kgosinkwe or Nciza to display their vocal talents. The song that would represent a shift in their sound and point toward much of their future success was the hit "Majika." Retaining the kwaito beat, the song was more structurally tight than their previous hits and more lyrically dense; it borrowed heavily from the sound of already-established pop artists such as Brenda Fassie and Ringo. The song became a huge hit over the 2000 festive season. It also became a favorite at wedding parties in the year after it was released. It successfully combined elements of neotraditional music with house-backed kwaito inflections.

FIGURE 6.3. Mafikizolo, *Gate Crashers* album cover (2005). Source: Kalawa Jazmee.

The video for the song "Majika" represents one of the important ways in which Mafikizolo manipulates notions of tradition, redefining what constitutes tradition for their own purposes along the way. Tradition itself is a type of historical trope, a reified concept that can be strategically deployed and manipulated to delineate between a past and a present. When the presence of traditional South African cultural features, such as sangomas and traditional attire, can be observed on the contemporary landscape, they index a pre-Western influenced history, and a pre-apartheid history in which such elements had much more symbolic and social value. Accessing and manipulating "tradition," then, becomes a way for contemporary kwaito artists to mediate between history and the future.

Perhaps in recognition of the fact that the sonic texture of the song was a departure from the typical kwaito fare they had offered up on the *Gate Crashers* CD, the group decided to film the video for "Majika" using drastically different sartorial and visual regimes from that of their earlier videos. "Majika," filmed in a replica traditional village, features the group members of Mafikizolo dressed in various ethnic attire, reenacting a traditional Zulu wedding.[5] Mafu wears the colorful yet cumbersome Ndebele women's attire. Throughout much of the video, Kgosinkwe and Madingoane flank her, dressed in Xhosa and Zulu attire, respectively. While the song's lyrics are in Zulu, and the bride and groom are attired in traditional Zulu dress, the three members of Mafikizolo, as well as other wedding guests, are not dressed in Zulu attire solely. Furthermore, the traditional attire choices by the members of Mafikizolo do not necessarily correspond to their own ethnic backgrounds. Filmed in a replica, rather than an actual, village in KwaZulu-Natal, Mafikizolo acknowledges the artifice of their brand of tradition, particularly the fact that it can only be enacted through modern technology. Midway through the video, the group does a number of Congolese-influenced dances that became popular in the late 1990s and early 2000s in South African kwaito performance and begins the traditional wedding processional; then the beat of "Majika" is interrupted by the beat of the group's previous hit, "Lotto." This interruption literally freezes the activity of the video; the dancers are stopped mid-step. This moment of interruption serves two purposes. First, it reminds the audience that the song, for all its sonic and visual dissonance with their previous hits, is still a Mafikizolo song by tying "Lotto" to "Majika." Second, the moment of interruption also serves to reveal the constructed and unstable nature of the tradition that Mafikizolo accesses. The performance of tradition draws from a set of practices that are shifting and, with the insertion of the beat of "Lotto," ephemeral. The moment of disruption, the injection of kwaito into tradition, reveals kwaito bodies to be anchored to a set of musical and cultural traditions that precede it. Madingoane resolves this disruption by going over to the radio and switching the station, the sound picking up from where it was left off, the activity of the video continuing as if the moment of interruption did not occur. However, it might be equally useful to read the moment not as a full resolution, as kwaito being put back in its place, removed from the space of tradition and the village, but instead as a moment of disjuncture (Appadurai 1996). The relationship between the past and the present is unresolved and unknowable, informed by global cultural flows even as it responds to local needs and concerns. Thus just as tradition produces kwaito, kwaito produces the tradition, and both are part of global flows and exchanges.

While the video features some traditional Zulu dances, none of which are performed by the actual group members, the bulk of the dancing focuses on contemporary popular kwaito dances of the time period. As such, a deliberate disconnect is created between the ostensibly traditional setting and attire and the contemporary dances and music. However, if we understand the video as a demonstration of forms of historical disjuncture, we can read the video as a performance not of disconnect but of deliberate polyphony where seemingly disparate elements reveal a coherence in their own right. The video for "Majika" reveals a tradition that is formed from encounters that are urban, syncretic, and ethnically mixed (see chapter 2). One does not have to be Zulu to do Zulu dances, wear Zulu traditional attire, or sing in Zulu. In fact, contemporary kwaito dances in this particular setting become at this moment as traditional as anything else. Eyerman and Jamison (1998) indicate the importance of notions of tradition to the concept of musical recall, reminding us that tradition "in its periodic recombinations and reformulations" is key to understanding the relationship between cultural practices and politics (77). This is because, as they describe, tradition is "the past in the present . . . vital to our understanding and interpretation of who we are and what we are meant to do" (161).

The popularity and success of "Majika" prompted Mafikizolo to consider increasing the incorporation of older, more neotraditional sounds of South African popular music into their brand of kwaito. While it is important to emphasize that other kwaito groups had experimented with incorporating neotraditional songs into their act, these songs typically represented only a minority of the songs on an album, and few artists built an image around neotraditional styling. Beginning in the early 2000s, Mafikizolo completely changed their image, rebranded themselves, and made the incorporation of older musical styles their hallmark. The new tracks on their album *Sibongile* (2002) relied on samples from 1950s-era Afro-jazz, particularly *marabi* music.[6] In addition, the group seemed to be capitalizing on the explosion of Sophiatown in South African popular culture. Numerous *kofifi* parties emerged throughout South Africa, as people were encouraged to relive the "good times" of shebeen culture by dressing in Sophiatown-inspired attire.[7] Fashion houses, such as the Johannesburg-based Stoned Cherrie and the Cape Town–based Sun Goddess, also captured the feel of Sophiatown and garnered much success with their vintage-inspired clothing.

Mafikizolo reworked neotraditional and South African jazz music of the past (marabi, kwela) into a remembrance of and celebration of Sophiatown aesthetics and ethics (see figure 6.4). While the success of "Majika" may have been somewhat unexpected, the subsequent shift of their style of music to incorporate additional Afro-jazz and neotraditional elements styled through Sophiatown was a deliberate act that played on the significance and meaning of Sophiatown in contemporary South Africa. Sonically, this represents an example of post-memory that forms the basis of a societal musical memory. Brassy horn riffs and prominent bass playing became key fixtures in the sampling of their music, creating a hybrid sound that blended the electronic styling of kwaito with Afro-jazz. According to Kabir, post-memory is a central component to musical recall. Kabir draws from Marianne Hirsch, who describes post-memory as "the experience of those who grow up dominated by narratives that

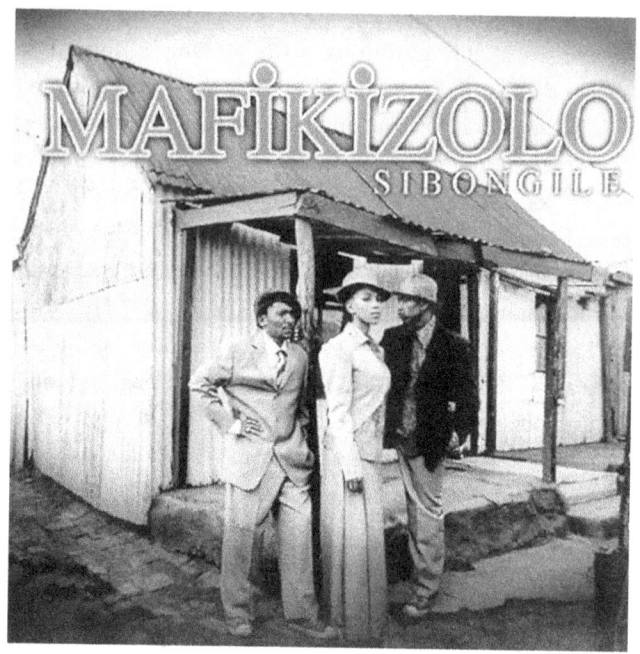

FIGURE 6.4. The cover for the "Sibongile" single (2005), reminiscent of Sophiatown, featuring the original Mafikizolo members (*left to right*): Tebogo Madingoane, Nhlanhla Nciza, and Theo Kgosinkwe. Source: Universal Music Group.

preceded their birth, whose own belated stories are displaced by the stories of the previous generation, shaped by traumatic events that they can neither understand nor create. . . . [It is] a space of remembrance, more broadly available through cultural and public, and not merely individual and personal, acts of remembrance, identification and projection" (Hirsch, in Kabir 2004, 172). Post-memory refers to the younger generation of South Africans who grow up with the stories of the past. Pain and cultural trauma are often communicated through the music of that era. While parents and grandparents may deliberately choose to not mention a painful past, music of the Sophiatown era preserved what official apartheid history books and destroyed buildings could not. Thus, younger members of the Black community grow up with knowledge about events that they may have never seen or experienced for themselves. This form of memory affects many of the citation practices involved in kwaito musical practices, particularly in relationship to Sophiatown.

The success of "Majika" and the mining of history and tradition spurred Mafikizolo to explore further this style of mixed kwaito, which incorporated additional musical elements from the past. Significantly, the revival of Sophiatown was as much a marketing phenomenon as a political one, and these two aspects of ensuring the remembrance of Sophiatown in public discourse are not disconnected. This does not mean that the remembrance of Sophiatown can be easily reduced to commercial manipulation; however, it does speak to the ability of marketers to recognize the affective power that Sophiatown holds over the consciousness of post-apartheid South Africa. This affective relationship was to be very important to the success of Mafikizolo. However, the desire to think about Sophiatown in relationship to their music and image was not simply a commercial one. Speaking of their album *Sibongile*, Nciza explains:

> After we recorded the album we sat down and we were trying to figure out what the theme of this album was going to be so we decided on the fifties theme. It was influenced by the first track on the album which is called "Marabi." Marabi is basically the music from the fifties in South Africa where you had your Sophiatown and we also went back to our history and we wanted to celebrate the fifties artists and the fifties music and just bring it back to today because a lot of people our age don't know what used to happen in the fifties, what kind of image they used to have, the style of clothing, the style of music. We believe that the fifties musicians contributed a lot into the South African musical heritage so we sat down and came up with this tribute to the fifties period. You can see on the cover of the album we are in our fifties outfits. (Nciza, in Kaganof 2006a)

Here, Nciza positions Mafikizolo as both inheritors of a tradition worth preserving and remembering and educators to the current generation. Nciza explicitly frames Mafikizolo's incorporation of Sophiatown aesthetics as one of remembrance for a musical style and sartorial practice centered on marabi. While she is not clear on why this particular juncture in South African history was chosen, I contend that the memory of Sophiatown as a place of Black independence, success, and cultural creativity under tumultuous circumstances informs much of why its memory is so powerful in the post-apartheid period.

Mafikizolo's four albums (*Sibongile*, *Kwela*, *Van Toeka Af*, and *Six Mabone*) in which they deployed the Sophiatown theme have all been major successes. As Nciza suggests, the visuality of the Sophiatown image was an extremely important part of making the connection between contemporary and historical periods. Fashion was central to how the group performed their interpretation of Sophiatown. Mastering the art of Sophiatown sartorial practices was necessary in order to remaster 1950s fashion for a new generation. *Sibongile* was the first album to feature the group in Sophiatown attire. To signify the historical aspects of the imagery, the album cover is sepia toned. The group does not simply imitate the Sophiatown style. Instead, like much of the Sophiatown-inspired fashion of Stoned Cherrie, the group plays with and updates the styles of the period. Madingoane probably keeps the closest to a traditional Sophiatown look, posing in an oversized suit from the period with a matching beret. Kgosinkwe eschews the suit for an oversized sport jacket but includes a fedora, and Nciza wears a long skirt but matches it with a button-down shirt and cropped long-sleeve blazer along with a fedora.

The recuperation of 1950s style central to Mafikizolo's public presentation was evident in the growth of a bold new fashion industry that was increasingly led by young Black designers. Speaking specifically of Stoned Cherrie's style, Christian M. Rogerson, writing about the contemporary fashion industry in South Africa, argues that "what has emerged is a range of designs which are inspired by urban township culture and a 'cheeky definition of what it means to be African.' Stoned Cherrie blends images of boxing champions, beauty queens, and musicians from *Drum*, a magazine that was extremely popular in the 1950s, integrating them into contemporary fashion styles" (2006, 230). Importantly, the remixing of Sophiatown (Fink 2014, 286) relies on a "conglomeration of medially (re)produced images, sounds, and texts" that creates performative memory. In order to re-create the performance of Sophiatown, in order to enact its musical memory, Mafikizolo would have had to access *Drum* archives, raid the closets and memories of parents and grandparents, watch films or television programs about the "*Drum* decade," examine old photos of family members, listen to old music, and shop at thrift stores and contemporary fashion houses such as Stoned Cherrie, which itself relied on mediated images to create the look. Equally important is the Afrodiasporic nature of the Sophiatown look, which "were hybrids themselves, having been derived from 1950s American styles" (Farber 2010, 151).

The themes of Sophiatown continued on subsequent albums, which all relied on well-known phrases or words from Tsotsitaal (the language of Sophiatown) to convey the Sophiatown image. *Kwela* refers to the Afro-jazz of the 1950s, but it also refers to the police vehicle, called a kwela-kwela, that was used to round up individuals who had violated various apartheid laws restricting Black mobility. On the album cover for *Kwela*, in color, the group appears in more polished and sophisticated Sophiatown attire, the men in patterned jackets and fedora hats, Nciza in lace gloves and a long, flowing white dress (see figure 6.5). Their next album, *Van Toeka Af*, plays on a well-known phrase of 1950s Sophiatown, used to express something that either has a long history or is from the distant past. In the case of Mafikizolo, that distant past was barely a half-century old. The first album since the tragic slaying of Madin-

FIGURE 6.5.
Mafikizolo, *Kwela* cover (2005).
Source: Universal Music Group.

goane features the group, now a duo, dressed in more casual Sophiatown-inspired outfits featuring patterned sweaters and vests, rather than the suits and formal dresses worn in earlier photographs. The back of the CD cover shows the duo in wedding attire, paying homage to the importance of weddings and celebrations as central spaces for the dissemination of Mafikizolo's music and dances.

During the time in which I completed fieldwork, I had the opportunity to observe the ways in which Mafikizolo's music was incorporated into weddings and similar types of celebrations. While attending a wedding in Soweto, I was told that there was going to be a pairing off of couples dancing in a wedding processual to welcome the bride and groom to the reception tent. They had married earlier in the day, gone to the Moroka dam to take pictures (a popular spot for wedding pictures in Soweto), and were on their way to the reception, which was being held on the streets of Soweto underneath a tent. My friend stated that I could be his partner in the dance (in an interesting disruption of hetero [if not homo] normativity). The song that played as we danced in two lines (men on one side, women on the other) and performed stripped-down versions of the dances displayed in the video was Mafikizolo's "Ndihamba Nawe" in which the principal singers (Nciza first, followed by Kgosinkwe) profess their desire to be with one another ("Ndihamba Nawe" roughly translates as "I choose you"). Along with the easy-to-perform dance moves that could be readily mastered by anyone from a young child to a grandmother, the lack of profanity or deep slang, and the retro-pop sonic styling, it became clear why Mafikizolo's music was a staple at wedding parties. The songs told the story of heterosexual coupling and provided a sonic and kinesthetic narrative that was cross-generational and accessible.

Their final album featuring the Sophiatown theme (and the last before the group went on a hiatus to pursue solo projects) was entitled *Six Mabone*, in reference to a type of car, which was considered a flashy desirable vehicle in its heyday. According to the group members, the Six Mabone was as coveted in its day as the BMW or Mercedes is among the contemporary Black elite. The album cover features Kgosinkwe and Nciza dressed to the nines, leaning on a Six Mabone. Nciza wears a gold lamé dress, a fur stole, and diamond jewelry around her neck; Kgosinkwe wears black suit pants, a white button-down shirt, a black sweater vest, a tie, a fedora, and wingtip dance shoes. Other promotional material for the album features the duo at a piano, Kgosinkwe with "conked" hair, wearing a slightly oversized black suit with red stripes and a red bow tie. Nciza compliments him in a red dress, a diamond choker, and a fox fur on her shoulder, once again performing 1950s glamour.

When possible, the music videos for Mafikizolo reflect the Sophiatown theme. "Marabi," a song from the perspective of a mother warning her child about the dangers of Johannesburg, saw Nciza dressed in 1950s-inspired clothing by Stoned Cherrie, flanked by Kgosinkwe and Madingoane in suspenders, ties, and fedoras. While singing the lyrics, the group performs a number of popular dances from the Sophiatown era. The idea that Johannesburg might be a dangerous place for a new inhabitant is not limited to the past. Rather, the juxtaposition of images of contemporary Johannesburg combined with the group members' dancing and attire (in the video, Nciza wears a Stoned Cherrie *Drum* T-shirt) serves to remind the audience that while the song is inspired by past histories of migration, the sentiment is still as current as ever. This marabi-inflected song with lyrics that initially seem anachronistic was extremely popular, because even in contemporary South Africa, migration to Johannesburg (or other large urban conglomerations) and the hopes and fears about the success of such ventures are strong in the public consciousness. And while the deliberate separation of families that characterized apartheid migration is not legally enforced, individuals still have to leave families (often children and sometimes spouses or significant others) behind in order to make a living.

Another example of the use of Sophiatown themes can be found in the video for "Kwela." Featuring Hugh Masekela, the song chronicles the unfair arrests of Black citizens in apartheid South Africa under the pass law regime. Filmed inside a prison, the group members are taken by a white guard into the jail where mug shots are taken and they sing about the injustice of their arrest. The video is interspersed with archival images from 1950s Johannesburg, demonstrating Fink's (2014) point about both Sophiatown as a performative archive of history and the necessity of mediated images to the performance. Images include cars traveling on the M1 freeway, the construction of skyscrapers, and protest marches. Eventually Masekela joins the trio in prison, and the video concludes with a mini uprising as Masekela and Mafikizolo drive the white prison guard from their cell. Their dancing and singing become the basis through which they resist the dehumanizing conditions of apartheid.

However, it is probably in the realm of the sonic that Mafikizolo makes the largest impression. The first popular hit that was a Sophiatown-themed song was "Ndihamba Nawe," the lead single from *Sibongile*. The horn sample from Sophie Mgcina's "Mmangwane" was used to create the song's signature sound. Unfortunately, Mgcina was not initially credited for this sample, and the group eventually had to settle out of court with her for royalties as well as provide proper credit on future releases of the album. *Kwela* featured the aforementioned song "Kwela," in which Kgosinkwe sings in an infected style reminiscent of the guest artist Masekela, who lends his trumpet (and vocals) to the melody of the song. Yet the most popular song from *Kwela* proved to be inspired by Letta Mbulu's "Meet Me at the River." "Emlanjeni," a wistful love song that speaks of lovers separated for years, recalls the times when Black South Africans (particularly men) would migrate and see their families only once a year, if that often. The song tapped into this sentiment of labor and separation from love, that, while less a reality today, still organizes affective relations concerning forced long-distance familial and romantic relationships. Many people find themselves going months without seeing loved ones because they cannot afford the transport costs or time off of work. As Nciza sings the opening plaintive cry of "It's been years since I have seen my boyfriend," she strikes an emotional chord both with those who live in these situations and those who have heard about them or experienced the separation of families as a routine aspect of everyday life or as an effect of post-memory.

Van Toeka Af includes a duet with Dorothy Masuka, "Sebenza," that is a kwela-influenced song replete with pennywhistle melodies. This collaboration with Masuka is an ideal departure point to discuss the gender and sexuality politics that inform Mafikizolo's performance of the past. Writing about the figure of the "woman" in representations of Sophiatown, the literary scholar Meg Samuelson notes that in written works from the Sophiatown era the woman "notably operates as an overly determined sign—marking the meaning of both rural and urban [space]" (2008, 64). Samuelson demonstrates how gendered representations in Sophiatown rested on a reformulation of urban women's bodies away from the notion of depravity toward domesticity from moral degeneration to cultural fusion (64–65). Women in the Sophiatown narrative and its representations were often reduced to the dutiful homemaker on the one hand or the dangerous jazz singer on the other. However, as demonstrated by her analysis of short stories of the Sophiatown era, the femininity performed by the women of Sophiatown often could not be contained by men's efforts to restrict the modern woman to the performance of her domesticity (67). Samuelson suggests that by recuperating the political femininity of women such as Dorothy Masuka, South Africa can imagine not only alternative pasts but different presents and more equitable futures. She writes,

> Masuka—who has survived the bulldozers, and apartheid itself, and "can still sing" today—encapsulates some of the wild potential that [Can] Themba's women begin to unleash. . . . Dating from the same period as Themba's Modern Misses, "Nolishwa" is remarkable for its presentation of a masculinity that welcomes a transgressive femininity. The song is comprised of two voices: a censuring, communal voice that critiques Nolishwa for her pride, ridicules her for wearing trousers, and reports in horror that she has been seen with another man; and the voice of Nolishwa's boyfriend, who retorts: 'She stands on her own two strong feet and I love her the more for that. . . . I love her as she is.' . . . Masuka's presence in our present returns us to an urban past of still untapped potential in which the Modern Miss begins to interpellate the figure of a new man with whom to craft new urban worlds. (73)

However, in remembering Sophiatown, Mafikizolo forgets, or at the very least downplays, the wild potential of alternative gender and sexuality performances in favor of a bourgeois and respectable domesticity that is given a contemporary twist.

The notion that Mafikizolo performs "wedding songs" is buttressed by the fact that the group began to offer up this theme as central to the visual and sonic texture of their performances. According to *Afripop!* magazine, "up and down the African continent it was for Mafikizolo's wedding playlist staple track 'Ndihamba Nawe' that they were known and loved" (Okumu 2013). Recall the traditional wedding in the video for "Majika," which cemented the group's foray into Afro-pop. Subsequent to this song, Kgosinkwe and Nciza perform respectable heterosexuality with each other (despite the fact that both are married to other people), the first hints of which are seen in "Ndihamba Nawe," whose lyrics suggest the choosing of one another. Madingoane has little else to do except run interference or play sidekick in this display of heterosexual love. In the video, Kgosinkwe plays a photographer who seemingly shuns his model/girlfriend for the more alluring Nciza. "Udakwa Njalo," a major hit off of *Kwela*, played off the reputation of the group to produce wedding songs. In its video, two backup dancers from Mafikizolo get married while the members of Mafikizolo form the most prominent attendees of the wedding party. "Emlanjeni" suggests the reuniting of heterosexual lovers (we are not clear if they are married) after a long time. In its video Kgosinkwe departs the home with Madingoane in the passenger seat of an old VW Beetle. The amount of luggage on the top suggests the journey is long and far. Nciza is attached to the home, remaining behind and dutifully completing domestic labor such as laundry, while reading letters ostensibly sent by Kgosinkwe from afar. In a split screen, the song reaches its emotional climax where the two leads exclaim that they will meet one another at the river. Madingoane is dropped off by Kgosinkwe (upon their return from his work). Kgosinkwe then finds Nciza gathering water by the river (further cementing her domesticity and performance of appropriate womanhood) as he returns to his love. Far from being an uncontained and uncontrolled domestic subject, Nciza is consistently presented as the ideal wife/mother for Kgosinkwe's appropriate and hardworking masculinity. When issues of "trouble" or disruption occur to the idealized heterosexual communion, the visual and sonic texts often eschew the Sophiatown theme or deliberately uncouple Kgosinkwe and Nciza. For example, in the video for "Makhwapheni," Mafikizolo's humorous salute to the radio and television show *Cheaters*, Nciza has a conflict with a cheating boyfriend and his mistress, but the boyfriend is not played by Kgosinkwe.

Perhaps the epitome of bourgeois domesticity performed by Mafikizolo is the song "Mas'thokoze." The song is the ideal wedding song, in that it promises wealth, eternal love, and happiness between Kgosinkwe and Nciza, who play the appropriately married pair in the video. Departing from the Sophiatown theme, the pair updates the narrative for a post-apartheid version of domesticity, mirroring the ideology that is promoted by magazines such as the contemporary versions of *Drum* and *True Love*. The pair presents an idealized Black bourgeois domesticity in which champagne is lavishly poured between friends, women are driven around by their lovers in luxury vehicles, and men are modern enough to cook for their women. This future-past sutures men's and women's bodies to appropriate masculinities and femininities in ways that belie the more complicated gender and sexuality performances of residents of Sophiatown. However, one aspect of Nciza's performance does consistently disturb the binary foundation through which the representation of gender relations is attached to Sophiatown. In the case of Nciza, she is both the alluring and potentially dangerous jazz woman and the domesticated housewife. That is, much of the Sophiatown performances enacted by Mafikizolo (whether live or in music videos or promotional material) center Nciza as a performer, dancer, singer, and worker detached from domestic duties. Her performances in the videos for "Ndihamba Nawe," "Kwela," and "Marabi," as well as the visual promotional tools for the album *Six Mabone* feature Nciza in the role of singer and working artist. She shifts between being the dutiful and domesticated partner of Kgosinkwe and being the entertainer who earns her living outside the home, in perhaps a less-than-respectable profession (even as she performs bourgeois respectable heterosexuality). Thus, Mafikizolo creates a gender and sexuality politics that renders Sophiatown and its remembrance as sanitized and acceptable in ways that are safe. The past that gets remembered is one where women, even when they worked outside the home, remained dutiful and attached to their domestic duties, were appropriately feminine, and could be contained by domesticity. This forgetting of Nciza's own complexity mirrors Pumla Dineo Gqola's (2007) critique of societal expectations for Black women, ostensibly powerful and in control at work, yet enclosed within the cult of femininity elsewhere. Commenting on the character Julie, played by Dolly Rathebe in *Jim Comes to Jo'burg*, Samuelson suggests perhaps the template of appropriate femininity that Nciza would perform to almost universal acclaim:

> But there is some cost to rendering Rathebe (and the city she represents) in the soft-focus light cast by *African Jim*'s happy dénouement. Lewis Nkosi, for instance, recalls that "she was much more explosive as a person in real life than she is made to behave in the film. It is as if you needed to tame those energies." . . . Indeed, the decorous Julie is far from the feisty Rathebe, who "embraced, embodied, and enacted the alternative moral order underpinning the Sophiatown imaginary." . . . As Lara Allen notes, "Rathebe was one of the first prominent personalities to reject publicly elite respectability." (Samuelson 2008, 65)

Mafikizolo's imagining of the past becomes a melancholic conviviality, one of domestic certainty and order in relationship to gender relations. Restless femininities, restless masculinities, and the alternative forms of sexualities are all expunged in their performances of Sophiatown. And yet even given the significant limitations in how gender, sexuality, and conviviality are imagined, there is still value in Mafikizolo's foray into musical memory for two reasons. First, Sophiatown is reimagined as a space of historical remembrance and re-invocation of "tradition." In engaging tradition, Mafikizolo reveals the unfixed categories of tradition and modernity. What they present is not the tradition of Zulu skins and rural, tribal identities. Instead, Sophiatown is recalled as a tradition of urban, modern African culture that is a blueprint for contemporary post-apartheid South Africa. This move to label the cultural production of that time period as essentially South African must always consider the Afrodiasporic dimensions of Sophiatown. These same Afrodiasporic dimensions continue to be at play in urban South African popular culture in the contemporary period, with kwaito being simply one example. Thus, Sophiatown and the politics of creating an urban Black South African identity are not disconnected from similarly situated contemporary debates and needs, and the struggle to define a modern, global, yet locally situated Black South African identity. Second, Nciza's kwaito body is always a threat to the normative femininities presented, since the combined role of disreputable jazz singer / respectable wife / mother contains within it the possibility for destabilizing the heteronormative ideal presented.

In their latest album, *Reunited* (2013), Mafikizolo shifts to a different time in South African history, claiming to be influenced visually not by Sophiatown and Afrodiasporic trends but by the "white 1950s and 1960s culture of roller skates and drive-ins and diners," although it might be worth noting that this "white 1950s and 1960s culture" was infused with significant Afrodiasporic (specifically African American) influences (Owen 2013). On the cover, the musicians appear as if they are in front of a classic Americana diner (see figures 6.6 and 6.7). A reporter had this to say about the shift in imagery: "My first thought is that they've lost the plot—Theo Kgosinkwe and Nhlanhla Nciza as John Travolta and Olivia Newton John doing the Grease thang? But later that night, when they revealed their radically different image, it did indeed work. Nhlanhla has a way of taking vintage fashion and molding it into her own. She looked young and fresh in her pink lipstick and blonde-cropped wig. It was all done with a sense of humor, a side to her she rarely shows in public" (Owen 2013).

The shift in visual imagery was connected to a significant shift in sound, which was more Afrodiasporic in mining West African and house beats. Syncopated rhythms and staccato phrasing blended together to create a more complex rhythmic timbre and a more upbeat song than Mafikizolo had engaged during their Sophiatown years. However, for Mafikizolo this return to kwaito/house, albeit with a West African twist, is almost a return to a future-past for them as a group as it represents a reappearance of their pre-*Sibongile*

FIGURE 6.6. Mafikizolo, *Reunited* cover (2013). Source: Universal Music South Africa.

FIGURE 6.7. The current members of Mafikizolo. Source: Universal Music South Africa.

house-inflected style, which spawned hits like "Lotto." This shift to a West African–influenced sound helped the song "Khona" become a pan-African hit, with Mafikizolo becoming one of the few kwaito stars to have continent-wide hits. With Nigerians and Ghanaians rushing to translate and interpret the Zulu lyrics, Mafikizolo has with their sound rerouted Afrodiasporic polyphony through the landscapes of West Africa, the American Midwest, and Johannesburg.

In the process, Mafikizolo also seems to have distanced themselves from the need to perform as a heterosexual couple married to idealized bourgeois domesticity. In fact, they use this album to queer memory. The video for "Khona" features the queer Vintage Cru dance team, a crew that has made a name for itself incorporating vogueing and waacking style into their dance repertoire.[8] Through taking these African American queer performance idioms and repurposing them in the South African context, Vintage Cru has created conversations about gender and sexuality and shifted discourses about African masculinity. Mafikizolo have not commented on why they chose to feature Vintage Cru so prominently in the video. However, given that "Khona" was such an Afrodiasporic hit, the group has presented an alternative version of African masculinity and sexual performance at a time when issues of queer rights have become politically charged Afrodiasporically.[9] Perusing YouTube comments, it is clear that the presence of Vintage Cru in the video sparked a number of debates about queerness and African identity. YouTube users Samuel Mwamba and bigboyvit mbiko argue against this presentation of alternative gender and sexuality, with Mwamba stating that "South Africa has a gay problem," and mbiko expressing annoyance that "too much [of] the Gay dancers at the beginning irritates our pure Black culture."[10] However, responses to these and other kinds of comments critiquing the inclusion of Vintage Cru suggest a healthy robust debate about the place of queer (male) sexuality on the continent. As user Tsholofelo Ranyane argues, "The rest of Africa has an 'institutionalizing the ostracizing/imprisoning/killing of people for their sexual preferences' problem. This makes me glad to be a South African living in South Africa :)," while user Boitumelo Mak states, "Actually South Africa's post-apartheid constitution was the first in the world to outlaw discrimination based on sexual orientation, and South Africa was the fifth country in the world, and the first in Africa, to legalize same-sex marriage . . . so yes that is exactly the sort of Freedom Nelson fought for. . . . [T]hat was his constitution."[11] Of course, as scholars and activists have demonstrated, constitutional protections for queer people in South Africa have not necessarily meant bodily freedom or sexual autonomy. However, in these responses what we do understand is that the complex relationship between the law and queer bodies in South Africa has created a sense that queer people (and Black queer people in particular) have claims to the body politic. In a shift from earlier representations, Mafikizolo presents a utopic future-past in which kwaito bodies articulate alternative gender and sexuality formations. We might site this shift for Mafikizolo in Nciza's body herself, as she inhabits the dual role of jazz singer / dutiful wife, her appropriately feminine performances containing the seeds of their own undoing. In reclaiming the unruly parts of Nciza's performance labor, Mafikizolo can resituate notions of tradition and history within alternative gender and sexuality formations. This is seen prominently in the visual and kinesthetic text of the video, where Nciza relishes the diva role in Afrofuturistic fashion, and also perhaps in Nciza's voice, which is deep, throaty, and at the lower end of her register.

In the video, contemporary chic and retro glamour combine with tradition to create a musical and visual mash-up, as Kgosinkwe and Nciza are the divo/diva surrounded by the alternatively white body-suited and gold-laméd Vintage Cru who vogue around them. The video accesses forms of tradition, through the members of Uhuru, and various Nguni-inflected dances, the visuality of Ndebele house painting, dashiki-clad dancers, and the prominence of a Zulu dancer, but like the video for "Majika," it is a blended urban tradition that draws from many influences temporally and spatially. It constantly remixes and recombines. The diasporic influences of the album *Reunited* draw as consistently from West African musical traditions as it does from house, creating an Afrodiasporic flow of possibility that returns Mafikizolo more centrally to the fold of kwaito and house and perhaps gives them the freedom to perform gender and sexuality differently. As one commentator writing about the video states, "Their [Mafikizolo's] decision to place them [Vintage Cru] so prominently could still be seen as a potentially dangerous idea on a continent where homosexuality is illegal in 34 countries. The acceptance of difference in a traditional setting isn't just bold; it speaks to the reality of plurality—as does this video" (Godinho 2013). As I discuss elsewhere (Livermon 2015), queerness and tradition are hardly mutually exclusive, and Mafikizolo has never attempted to present tradition or the performance of the past as linear or unconstructed. Yet in positioning Vintage Cru so prominently, Mafikizolo does boldly suggest (in contrast to their performance of Sophiatown) that queerness is a part of tradition, a part of the past that must be remembered, and a part of the future that should be constructed.

Even as Mafikizolo shifted gears sonically and aesthetically, the video text for "Khona" does relate to the glorification of Sophiatown in one important way. In creating a vision of an ethnically, nationally, and sexually diverse South Africa, combining traditions, sounds, and aesthetic references, the group presents an idealized post-apartheid harmony that harkens back to the mythology of Sophiatown. Racial mixture is perhaps the only kind of encounter not celebrated or visually encoded in the video. To this extent, the video text is another example of motswako as a theory or lens through which to think about encounter in South Africa. It is also, however, another reminder of why Sophiatown's mythic example of interracial and intraracial harmony retains its salience in the contemporary moment as the schisms that mark South African society reemerge into social movements and political programs that challenge the negotiated settlement between Black political organizations (namely the ANC) and the National Party.

School Uniform Parties, June 16, and the Remastery of the Suffering Black Body

I learned about my first June 16 school uniform party in 2002 while visiting a friend in Johannesburg. He and another friend were searching for just the right blazer to complement outfits that they had spent a couple of weeks preparing for. I got the sense that this party was quite a major social occasion and that the dress code for the party required attendees to wear school uniforms. At the time, I was completely unclear as to the symbolism of the uniform and only vaguely familiar with the history of June 16. I expressed some reservation about attending, given that I did not have a school uniform available and I did not know anyone who could lend me one for the evening. My friends reassured me, laughing and stating, "We'll just say you're an exchange student if anyone asks why you're not wearing a uniform . . . and in a way you are an exchange student." The preparations for the party seemed giddy and celebratory as my friends created looks that mimicked the school uniform worn by most South African students but were a twist on the typically conservative schoolboy look. Like myself, both of my friends were gay and they sought to wear their school uniforms with a dash of flamboyance and style. In the case of one of my friends, he ditched the pants that boys would typically wear and instead wore a pleated skirt. Colorful socks, ties repurposed as belts and bandanas, jewelry, and tennis shoes instead of dress shoes helped to repurpose the looks of my companions. All of these looks would have typically been "banned" had the wearers actually showed up to any school dressed in such a manner.

We drove to a rather upscale area of Soweto and parked our car. There were a lot of people at the party, and many of them dressed in various school uniforms.

While there certainly appeared to be a significant queer presence, there also seemed to be a mix of attendees. The street was cordoned off and revelers moved around in groups. In an attempt to replay the schoolboy and schoolgirl experience, tuck shops operated, selling candy, gum, chips, and *ama kip-kip*, and "students" could purchase *kotas* and sodas to drink.

Once the car was parked, we walked through the streets to the central location of the party. Importantly, the "walk" served as the runway, similar to our walk in the festival park, a liminal space that was a feature of all outdoor parties in which a great deal of the action takes place outside the official party space itself. I had managed to score a blazer from a friend of a friend and was able to put that on in my friend's car. I spied others changing, smoking, taking drinks, and making out in vehicles. As we walked closer to the crowds, we were also beckoned by the sound. This was the early 2000s and DJs were playing mostly kwaito mixed with house. The thumping beat of the four-on-the-floor sound that characterizes both genres could be felt as we got closer. As with most street parties, there was no central dance floor. Instead, the places where the crowd were most knotted served as the primary space for dancing; however, anywhere in the public space of the streets could turn into an impromptu dance party. The faux tuck shop operated from the house that served as the hosting space. In addition, speakers had been placed in the yard and just outside the yard of the host's home.

This was the era of kwaito's maturation and the sounds swayed with popular kwaito producer D-Rex's beats on songs like Mapaputsi's "Izinja," which trafficked in the tsotsi aesthetic, being about the "dogs" of the township. Replete with barks, an infectious easy-to-chant chorus, and the bassline-synth pairing popular during that time we got down to the business of dancing. The encounters enabled a queer-friendly space in which any possible pairing of intimacy seemed possible. My friends kept their eyes out for "boys" (guys who had the rough township performance they desired). But I remember this space particularly for the way in which it was able to engender a politics of queer women's performativity. The school uniform theme gave queer women much material to work with as school dresses and trousers could be worn in various different ways to accentuate the feminine form or allow multiple female masculinities to be displayed. At one point in the evening, our small group was incorporated into a group of queer women who exuberantly performed their sexuality, arms and bodies entangled as they swayed and twisted to the beats.

In my fieldnotes that I took that night when I returned home I wrote,

> just got back from this crazy party where everyone (except me) was wearing school uniforms. I saw a few celebrities including Theo from

Boom Shaka. Everyone was dancing and having a great time. It was one of the largest parties I have attended, I ran into some old friends I haven't seen in a while and saw an amazing mix of people, gay and straight having the time of their lives. Don't know quite what was going on or even who the host of the party was but I can only hope to be invited in the future. (June 17, 2002)

Later I was to find out that this school uniform party was one of the major events of the queer social scene in Johannesburg and many people looked forward to it each year. While not specifically queer, it was a space for Black queer Jozi and their friends to socialize, see, and be seen. While I never attended that particular school uniform party again, I did have the opportunity to attend a number of other ones over the course of the years. It was only later that I discovered that these school uniform parties held on June 16 were in fact quite controversial commemorations of a specific event.

A June 16 party, much like Sophiatown, is as much about the memory of the event and the performance of that memory as it is about the specific event itself. It is, like Sophiatown, a "performative archive" (Fink 2014) in which various contemporary stakes are at play, not the least of which is the contested issue over how Black pain and Black trauma should be enacted. Some general outlines concerning June 16 are necessary to contextualize what follows. Understanding the importance of language as a way to institute power, and sensing that Black South Africans were increasingly less likely to study Afrikaans, the apartheid government instituted the enforcement of Afrikaans (beginning in 1975) as the language of instruction in Black schools. Black students saw this move for what it was: a way to institute additional state power onto their bodies and a way to disconnect Black South Africans from an increasingly globalized world in which English was the medium of communication. English also helped to connect important global and Afrodiasporic resistance networks. As a result, students organized a number of walkouts and boycotts from schools during the fall of 1976. Students were organizing and resisting not only Afrikaans as a medium of instruction but also the larger contours of apartheid rule. The students from a number of high schools in Soweto (in particular, Morris Isaacson and Naledi high schools) had planned a boycott and peaceful protest and march against the enforcement of Afrikaans in the school curriculum. The police responded to these unarmed protesters with a great deal of violence and state terror, resulting in the deaths of a number of students. While the political activity of the students in Soweto is often spotlighted, it is important to note that the Soweto uprising, as it became popularly known, spawned a number of similar protests and resistance across the country among increasingly politicized Black students. The outcome of the protests was twofold. First, it reinvigorated an internal anti-apartheid movement that had been violently suppressed in the aftermath of significant political agitation by Black South Africans (and their allies) in the late 1950s and early 1960s. Second, it refocused Western attention back to the anti-apartheid movement at a time when the movement was being summarily dismissed as "communist" and therefore a threat to Western interests. While it may not have seemed so at the time, in retrospect, the protests of the students marked the beginning of the end for the apartheid regime, for the protests emboldened a new generation of anti-apartheid leadership that in the end made the costs of maintaining apartheid too high for the National Party government.

This is why perhaps the memory of June 16 has become particularly controversial in post-apartheid South Africa. The stakes concerning how that memory has been performed have turned into a moral panic about the state of contemporary youth. In the intervening years, one could not glance at a newspaper in South Africa without reading some account of how the commemoration of the Soweto uprising has been tarnished. Much of that criticism from everywhere on the political spectrum was leveled at the ways in which the ANC government either was failing the youth of South Africa or was attempting to monopolize the memory of those events for its own political gain. These criticisms, while important, are not the main focus of my examination here. Instead, my interest is in examining the ways in which Youth Day celebrations and the commentary about them become a "common sense" through which contemporary Black youth can be represented as in crisis.[12] Like the earlier discussion of consumerism and izikhothane, this moral panic serves to construct "Black township youth" and to suggest that these youth are lost and in need of saving. Unlike the discussion of izikhothane, the problem does not lie with corrupt Black politicians or the Black bourgeois elite who provide bad examples to the Black poor. Instead it lies within Black township youth themselves, who must in turn be politicized and reformed, lest their excessive and apathetic ways portend doom for a future South Africa.

The examples of the moral panic that surrounds the celebration of June 16 are numerous but they tend to concern drugs, drinking, and the ill effects (including sexual promiscuity and unprotected sex leading to HIV infection) that come from substance abuse. Comments from the Congress of South African Students are typical of the kinds of panic created during Youth Day celebrations: "our day is being turned into an official drinking day for young people and performing artists" ("Don't Drink on Youth Day" 2003). At Youth Day celebrations in 2013, the president of South Africa, Jacob Zuma, warned that booze and drugs were now the enemy of the youth: "As we speak today, many parents are in pain as they watch their children deteriorating and their lives being destroyed by drugs and alcohol abuse.... I have heard tales of children as young as eight who are now addicted to drugs. I have heard tales of young girls molested in drug dens.... [T]he youth have become slaves of drugs... others are slaves to alcohol abuse" (Zuma, in Hans 2013). Notice how in this speech there is no recognition of the structural forces that can influence young people's choices concerning the abuse of alcohol and/or drugs, including but not limited to underresourced and poorly funded schools, an inequitable and racially skewed labor market, and a limited social welfare system. Structural concerns of the political economy are not the primary danger for young people. Rather, their poor decision making is to blame. Anecdotal tales of wild, out-of-control youth, drunk and high and exposed to sex, prematurely work to construct "Black township youth" as being in need of saving, with those who represent major political parties all in possession of the correct prescription, which combines equal doses of limited government beneficence and personal responsibility. Eight-year-olds high on drugs and young girls (note the gendered nature of this construction) molested in drug dens create the representation of a nihilistic youth culture.

The need to control and manipulate youth behavior, particularly young women's behavior, is central to these particular panics. On the one hand, certainly young women are especially vulnerable to forms of sexual abuse within the country. But in creating the figure of the girl molested in a drug den as a warning to young Black South Africans about the dangers of excess, Zuma frames the problem of sexual abuse bizarrely. In this instance, the sexual abuse victim (a young girl) is blamed for her own victimization, her drug use placing her in a position to be violated in the first place. In constructing sexual abuse as the possible problem of the drug-addicted excessive girl bodies, Zuma's admonishments obscure the fact that most sexual abuse occurs in or near the home.[13] Even if marked by excess, like partying, drink, and drugs, this should not mean that victims of abuse are somehow culpable in their abuse. In constructing the (gendered) youth in this manner, they become the bodies in need of saving and the bodies that cannot be saved. That is, they are constructed simultaneously as hopeful and hopeless: hopeful in the sense that government programs (as well as programs created by corporations and nongovernmental organizations) can create societal change; hopeless in that the failure of these programs to create systemic change points not to structural deficiencies but rather to questionable moral choices on the part of "Black township youth." In this simultaneous construction, "Black township youth" are often talked about and talked to, but rarely talked with as one editorial commentator mentions: "if you preach to a teenager from a stage, they are not as inclined to listen to you. But if you speak to them on their level, chances are they might just hear and understand what you are saying" (Mahlerbe 2014).

In 2005, I was listening to a local radio station in Soweto when I overheard Winnie Madikezela-Mandela criticizing the youth of post-apartheid South Africa for turning June 16 into a day of drunken revelry. She opined that June 16 should be a day of somber reflection. Speaking specifically of Youth Day celebrations in 2001 and its connection with kwaito performers she stated, "The mere fact that the government needs kwaito artists to attract our youths to events like Youth Day shows you that we have lost mass appeal. Mandoza is now more popular than our leaders. . . . If I had been party to drafting policies around these very important days, I would have ensured [the days] are given the dignity they deserve. It is not African to celebrate death. In death, we pray, mourn, and sing solemn songs" (Dube 2001). Madikezela-Mandela constructs a number of arguments about the memorialization of the Soweto uprisings of June 16 to suggest appropriate ways of performing memory. While her words were specifically in relation to how the ANC government misappropriates the holiday and encourages a celebratory rather than commemorative tone, her criticisms access a larger concern about the appropriateness of how young people engage the memory of the apartheid struggle. Central to her argument is the notion, which she roots in a homogenous definition of "African culture," that death is to be honored through prayer, mourning, and the singing of solemn songs. Besides evoking rather essentialist and static notions of what should constitute African culture, what perhaps is most notable about Madikezela-Mandela's framing of the Soweto uprising is that it was primarily about Black death and that Black death has particular ways in which it must be performed and remembered. For Madikezela-Mandela, pain, trauma, and death enacted on the Black body means that celebration and commemoration are diametrically opposed binaries that besmirch not only the memory of June 16 but also the representation of African culture. However, two aspects of her reading should be contested. First, June 16, while certainly about "death" and the violence enacted by the apartheid state on the Black body, was not about Black death only. It was also about defiance, resistance, and the forms of Black joy, pleasure, and erotics that come from resistance. To reduce struggle simply to death and pain is to miss the full humanity of the moment. It is often in moments of struggle that other registers of feeling emerge, such as love, pleasure, humor, and excess. Second, her interpretation of "African culture" that suggests that death is only performed through solemnity leaves little space for solemnity and celebration to coexist (see figure 6.8).

Over the years, I have attended June 16 street bashes in Alexandra, Soweto, and Mmamelodi. I have been consistently struck by the fact that, in a country known for homophobic violence, particularly in the townships where these parties occurred, I saw a vast array of sexual identities and practices converging in a way that makes them queer. As one friend described it, "Straight-Sexuals, Homosexuals, Bisexuals, Asexuals, Metrosexuals all in one stylish party! [That is] the Soweto I know" (personal communication, 2013). The style that is on display is a remarkable testament to the ability to create and re-create in ways that blur the distinction between consumption and production. Lastly, the parties seem to be an instantiation of life, through the sheer joy on the participants' faces as they dance together, meet friends old and new, and exchange numbers for a variety of purposes ranging from the crudely sexual to the entrepreneurial.

In many ways, the school uniform parties are not distinctive in the vast array of contradiction, diversity, and difference that they are able to support in one space. And it seems to me that this form of celebration is in fact a way of commemoration. If nothing else, the generation of June 16 that rebelled against the apartheid government did so in order to secure an alternative future for South Africa, one in which the full humanity of Black South Africans can be realized. Instead of arguing that the school uniform parties of June 16 are somehow an affront to those who died on that day, an example of apoliticism, youth use these parties to enact a queer politics of memory through joy, celebration, and moments of excess. By dressing in school uniforms, dancing to their favorite kwaito and house musicians, and enacting a politics of pleasure in which a sheer array of diversity and creativity is made possible, they are engaging in the practices of life. These engagements in the practices of life are a key component to the vision of remastered freedom in South Africa. Through queering the memory of that day, the partygoers remaster its significance, changing Youth Day into a freedom dance that provides a space for alternative gender and sexual expression. The youth who struggled on that day paid an ultimate price for challenging the apartheid logic that consistently marked them as subhuman and killable. It seems to me that we do a disservice to their memory to only remember them as dead, traumatized, pained bodies. In literally redressing the day, the kwaito bodies of the post-apartheid era remember these dead, traumatized, pained bodies and remaster them into living testaments of the humanity that neither apartheid nor the post-apartheid period could ever fully circumscribe.

FIGURE 6.8. An example of a typical June 16 Youth Day party flyer. Source: Facebook: Ubuntu Kraac Beer Garden.

Conclusion

Kwaito bodies cultivate their own stance on historical events and processes and remaster them within the dynamic aesthetics of their art form for contemporary consumption. Processes of memory and historical refashioning not only serve to destabilize critiques that kwaito is simply tuneful music that is good for dancing and entertainment, but also create their own narratives about what freedom may look like as they seek to use sound, text, and performance to mediate between the past and our contemporary moment. The issue of memory, its relationship to Black trauma, and the performance of that memory is crucial for post-apartheid South African politics. The Truth and Reconciliation Commission in many ways centered on a retelling of the narrative of Black pain with the idea that the retelling and remembering of these moments would serve as an archive that would help the nation move forward into its post-apartheid future.

Two particularly fraught moments in history—the destruction of Sophiatown in the 1950s and the Soweto uprisings of 1976—engage this debate on how memory should be performed. Through an examination of Mafikizolo's use of Sophiatown imagery, the 1950s is remembered through forms of diverse conviviality, social order, certainty, and heteronormativity. In essence the image of a diverse, prosperous, heteronormative future-past is promised in ways that misrecognize the complexities (economic, sociosexual, spatial, to name just a few) of 1950s urban South Africa. The school uniform parties of June 16 and the moral panics about youth behavior reveal another moment of tension around the contemporary performance of the Black body. In this case, the complexity and contradiction of struggle is elided, and the youth of June 16 reduced only to their revolutionary political struggle, their sacrifice solely death, their commemoration only mourning.

Mafikizolo's "Khona" draws these two observations together, revealing the ways that queer remastery of traditions can refashion the past in alliance with the vernacular performances of June 16 parties. Musical memory through kwaito bodies imagines alternative traditions, creates spaces of conviviality, and imagines alternative futures. Some of these performances of memory participate in a cementing of heteronormativity while others imagine queer future-pasts. Yet it is this ability to account for difference and contradiction that informs the humanity of kwaito bodies and thus reveals the simultaneous possibilities and limitations of the post-apartheid moment. Within this space of contestation, the basis for locating a kwaito futurity and a remastered freedom is formed, one that is a direct challenge to the older Black male heteropatriarch as the primary change agent in contemporary South Africa.

Coda

Kwaito Futures, Remastered Freedoms

I return to the concept of remastery, particularly in relation to Keguro Macharia's (2016) call for new vernaculars of freedom. As I have argued, kwaito performance practices are one critical site through which we might come to fashion these new political vernaculars. Kwaito bodies use creative play and deep resolve to be who they are, finding pleasure in the fraught everyday of post-apartheid urban life. In the process they offer alternative ways to consider what freedom might look like and which practices might sustain it. Through the remastery of space and subjectivity, kwaito bodies push the limits of freedom, re-creating the parameters of what is possible. Their embodied practices point to the alternative ethics of freedom that Macharia suggests is so urgent in contemporary Africa. Here, I briefly examine three contemporary eruptions of kwaito political vernaculars that reveal an ethics of freedom and the continued relevance of the work of kwaito bodies.

Kwaito Futurity I: Pussy Parties

Toward the end of my fieldwork, Braamfontein was in the process of inner-city redevelopment and regeneration. This process framed Braamfontein as a nightlife and entertainment district. In the contemporary moment, this process of gentrification has accelerated, and the area has become populated with fashionable cafés, art spaces, restaurants, and trendy apartments in addition to its more historical function as a student-centered neighborhood catering to attendees of Wits University. A number of new bars and clubs have opened in the area, predominantly serving students and a hip, young, fashionable crowd. Much of the energy and eclecticism that I identified in early and mid-2000s Melville seems to have migrated at least temporarily to the Braamfontein night scene. One of the most interesting elements of that night scene is the monthly pussy party.

Started by two women artists and DJs, the Soweto-based DJ Phatstoki (Gonste More) and Rosie Parade (Colleen Balchin), the monthly party is envisioned as a "progressive femme-friendly event in Johannesburg that hopes to carve out a safe space where women and femmes can express and celebrate their femininity in whichever way they want" (Bogatsu 2016). The DJs (one Black, one white) aspire for the space to cultivate femme-centric joy by celebrating women and femininity in various performing bodies. Also key to this practice is the promotion of women DJs and a culture of consent and safety in nightlife space. According to Rosie Parade, she was surprised at the level of panic the word "pussy" inspires: "when guys say pussy they're talking heavy shit." Phatstoki elaborates, "They [men] oversexualize the word . . . at any point you take ownership of something they use to disempower you, then it becomes a problem" (Bogatsu 2016). The re-appropriation of "pussy" suggests the type of femme remastering of language that might inform new practices of freedom that at the very least are genealogically connected to earlier feminine kwaito performances that sang freedom in a new register: "[the party is] building on the legacies of Brenda Fassie and Lebo Mathosa" (Bogatsu 2016).

Rosie Parade sees the pussy party as an extension of her artistic work, particularly her DJ-ing. She had this to say about the effect of music and how dancing to it creates a kind of power that is freeing for the body: "It's the silences between sounds that allow us to find the beauty in rhythm and beauty is power and smile is its sword" (*The Fuss* 2016). The smile as a weapon of joy, release, and freedom on the dance floor is part of the ethos of the pussy parties. It is a way to also carve out a tangible space apart from the masculine and male-dominated music scene of South Africa. Also part of the politics of the space is

FIGURE C.1. An example of DJ Phatstoki's self-created pussy party invitation flyer. Source: DJ Phatstoki.

introductory DJ-ing workshops specifically for young women. These new DJS are then invited to play a set at a future pussy party to hone their skills. "While each party still has a surprising number of guys in attendance [who all have to pay an entrance fee of an amount left to the discretion of the woman at the door], the event is unlike anything else in Johannesburg. Picture a pro-femme community of artists, activists, and friends moving around carelessly to experimental sounds" (Bogatsu 2016) (see figure C.1).

The pussy parties are one eruption of an alternative political vernacular centered on the creation of women and femme-centric space. They see themselves building on the celebration of Black women's erotic energy that traces its lineage locally to Fassie and Mathosa but also engages a host of other women working Afrodiasporically through sound and performance, including the New York–based (of Trinidadian and Jamaican descent) underground queer hip-hop performer JUNGLEPUSSY (Shayna McHayle) and the Bronx-bred, Trinidadian-Dominican stripper-turned-rapper Cardi B (Belcalis Almánzar). If the political vernacular of post-apartheid South Africa worked from

FIGURE C.2. Babes Wodumo. Source: West Ink.

the intentional spaces of pussy-centricity, how staggeringly different might the alternative political futures of South Africa look and sound? Perhaps the nation might step away from a limited debate about incorporating Black women into already existing hierarchical structures and turn its attention instead to remastering those structures.

Kwaito Futurity II: Babes Wodumo and Ratchet Politics

Bongekile Simelane, known professionally as Babes Wodumo (literally, the famous Babes) is a Durban-based young Black woman dancer and music artist who in the last few years has captivated South Africa with her infectious dance songs. Her 2016 song "Wololo" was considered by many to be the anthem of the year. Operating in the musical style of g'qom, which is considered by many to be a not-so-distant relation of kwaito, Wodumo has emerged as one of the few women figures in the genre. Wodumo is important because she has been cited as inheriting the legacies of Fassie and Mathosa as a contemporary empowering example of "women and femme artists inspiring sexual liberation"

(Bogatsu 2016). However, what I find most compelling about her has been the ways in which she, her fans, and her detractors engage the discourse of "ratchetness" to describe her persona and public performances (see figure C.2).

As a particular mid-2010s formation, "ratchetness," though originating in Southern African American vernaculars, has developed Afrodiasporically to describe a particular remove from a politics of propriety and respectability. It has also implicitly and explicitly focused on the behaviors and actions of working-class Black women. "Ratchet has become an umbrella term for all things associated with the linguistic, stylistic, and cultural practices witnessed or otherwise of poor people, specifically poor people of color and more specifically poor women of color. . . . [R]atchet is a very feminine gendered term" (Sesali 2013). In creating an alternative space of performance and being for Black women, several scholars have argued that ratchet is a form of liberatory practice that evinces a "shadowy" underground feminism (Stallings 2013, 138). According to therapist, creator, and Black queer troublemaker Montinique McEachern (2017), it is in spaces of what she calls "anti-respectability" that she is able to situate practices of freedom as a Black woman. "Ratchet is the embodiment of Black Femme liberatory discourse. Ratchet is a cultural knowledge performance and awareness of an anti-respectability that can be shared across Black communities" (79). Importantly, McEachern identifies musical contexts as a key site of diasporic ratchet feminism: "ratchet feminism is created across contexts that Black girls are in and can be seen in elements of dancehall, soca, and other diasporic music genres" (80). In the South African context, we might read the inclusion of ratchet into a local vocabulary as being situated in similar gendered-racial politics that frame Afrodiasporic Space as a site of performance and exchange.

In numerous interviews, Wodumo has described herself as ratchet. Her displaced weave in multiple bright colors that hangs precariously on her head, her dance styles inspired by the township shebeen, her not always perfectly manicured nails, her insufficient command of English, her crop tops, and her lyrics all mark her in the South African consciousness as ratchet (Lindwa 2016). In claiming her performances as ratchet, we might ask what practices of freedom she is attempting to access. How might we link those practices to kwaito bodies and the new political vernaculars created through kwaito? In a recent article written about the hypocrisy of consuming the practices of working-class Black women in South Africa, writer and journalist Buhle Lindwa (2016) had this to say about Wodumo and her detractors: "many carefree ghetto Black girls have influenced South African pop culture, but never get the credit they deserve, and Babes is not the first." Importantly, Lindwa links Wodumo's eruptions of "disrespectability" to previous musical artists, including Fassie and Mathosa. She argues that when Black middle-class women adopt toned-down performances of the "ratchet ghetto girl," they are celebrated as carefree, yet working-class Black township women are derided for the same behavior. Lindwa makes the case that such an erasure and violent policing have a long history in South African public culture, and her goal is to give poor and working-class Black women the credit they deserve as cultural innovators and tastemakers. Interestingly, Wodumo herself attaches being ratchet to a performance of carefree-dom: "I grew up in a township, that's why I say I am ratchet. . . . [P]eople always think when I say I'm ratchet I mean I'm a hobo, I actually mean the carefree kind of ratchet" (Wodumo, in *The Edge* 2016). In this case, Wodumo enacts many of the possibilities of ratchet feminism, an eruption into public life of the carefree, I-don't-give-a-fuck, fearless Black girl from the ghetto into popular culture. Central to the genealogy of Black women's performance that Wodumo inherits is a shamelessly confident claiming of space.

In imagining kwaito futurities, what makes Wodumo an interesting emergent figure is precisely the fact that she performs a type of femininity unapologetically rooted in township culture and not seeking rehabilitation. We might consider her femme ghetto-centricity an important counterpoint to the forms of rehabilitation so central to the hypermasculine tsotsi performativity as inhabited by Mandoza. In essence, her radical femme performances undo the masculine thug as the (ir)redeemable subject. The death of Lebo Mathosa did leave a vacuum in the performance of dangerous women in post-apartheid South Africa. While it is unclear precisely which direction Wodumo's public life will take, she has served as an apt inheritor of Black women's ratchet politics. If the pussy parties use the legacy of kwaito performance to carve out re-imagined cultural spaces that center women, Wodumo draws on the legacy of kwaito as ghetto/township music and the gendered possibilities within it to remind us of the racialized, classed, and gendered dimensions of a kwaito futurity. If kwaito is salient for nothing else, it is as a cultural space for the eruption of young Black township culture into the public consciousness of South Africa. Wodumo's insistence on her ratchet performances mirrors the potential classed disruption inherent in kwaito. The transition to a heteropatriarchal post-apartheid future assumes that the economic frame of South Africa would undergo little change, only that upper- and middle-class Black people would be admitted. There was an assumption (not completely inaccurate) that the higher-class Black person would also carry a veneer of white cultural capital with them. Part of the dissonance created by kwaito and the tensions inherent around someone like Wodumo having access to capital is precisely that ratchetness refuses this white upper- and middle-class respectability as a condition for inclusion. Furthermore, as a feminine-gendered ghetto performance, it is not available for nationalist and corporate recuperation (Lindwa 2016).

Kwaito Futurity III: FAKA and Siyakaka Feminism

Founded by Buyani Duma (Desire Marea) and Thato Ramaisa (Fela Gucci), two young Black gender-nonconforming queer men, FAKA began as a performance art duo that has branched out into a variety of media. They now consider what they do part of a cultural movement, one that "represents an unapologetic representation of Black Queer culture in South Africa that embraces the intersections of race, gender, sexuality, and spirituality" (FAKA n.d.). They see the various platforms that they engage—which range from art gallery performances to SoundCloud mixes to their own curated website—as "an archival project documenting young Black queer creative's" who do not receive "the exposure and recognition they deserve" (FAKA, in Leiman 2015). As part of their cultural movement, they emphasize "sexual fluidity and othered masculine identities," pushing the boundaries of representations of Black masculinities in the South African public sphere. As a group, they specifically cite kwaito pioneers Boom Shaka as part of the performative foundation that they are building on. For them, kwaito is a space of inspiration, where young, Black creatives "were celebrated for the integrity with which they expressed their identity and desires. It's [FAKA] really a continuation of that" (Leiman 2015). While there are many aspects of their performances that we could consider to be forms of kwaito futurity, I focus here on the concept of Siyakaka feminism, which permeates their artistic production (see figure C.3).

According to Gucci, "faka" in Zulu means to penetrate or to occupy; he and Marea conceptualized FAKA as a way to occupy spaces that exclude Black queer people in South Africa ("Desire Marea: Siyakaka"). Siyakaka feminism is a manifesto that works to mend what is broken within Black South Africans and transcend all falsified binaries. Its inspiration comes partially from a quote attributed to Nelson Mandela, concerning the artists Boom Shaka. Marea said that, while growing up, they did not like Mandela because they were told at a young age that Mandela disapproved of Boom Shaka. Allegedly, when asked about their risqué outfits and dance moves, Mandela replied, "Mina, andizifuni eziz' Boom Shaka," meaning, "Me, I do not want these Boom Shakas" ("Desire Marea: Siyakaka"). Marea recounts the disarticulating affect that they

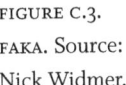
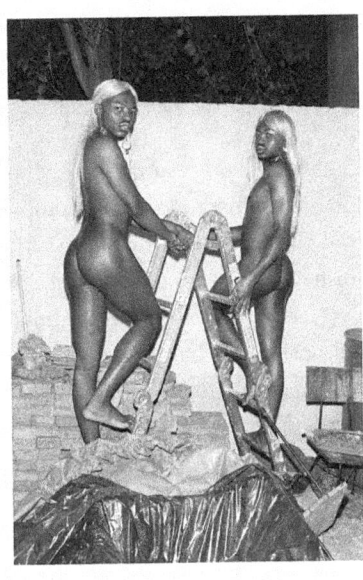

FIGURE C.3. FAKA. Source: Nick Widmer.

felt for the post-apartheid project when they discovered that the most important figure of liberation disapproved of Boom Shaka, whose prominently featured women members were, for them, beautiful. Evidently, Mandela did not have a taste for the way the women of Boom Shaka represented themselves; perhaps he found their bodies and their performance disruptive to his values. Marea recalls how this attributed quote was then used to police their body as a young femme boy who loved dancing to Boom Shaka and imitating the moves of Lebo Mathosa and Thembi Seete. They remember the sense of pain and alienation that this form of policing produced.

This individual policing is part of a larger problem in which a set of cultural and institutional practices are consistently reinforced in a way that shames Black women in particular and Black femininity of various sorts more generally. Marea connects the policing of their own body with the larger struggle of Black women against their disproportionately violent oppression, policing, and disenfranchisement in post-apartheid society. Echoing my earlier discussion, Marea explains that what they most connect with in the performances of Boom Shaka are the politics of liberation that the women embodied as two Black women who owned their Blackness and their sexuality. This is a radical intervention in a society where Black women are taught that they have no right to own their sexuality or their bodies. For Marea, Boom Shaka represented an important break in societal heteropatriarchy and the fact that Mandela or any other major figure of South Africa's liberation movement could not recognize them as symbolic of a new post-apartheid freedom proved most disheartening for the future of South Africa.

Marea asks, Who are you to say how I should express myself in my body? How can you have an honest conversation about liberation or freedom without me? Does who I am exclude me from the future of the nation? Marea refuses any post-apartheid future that refuses them; it can only produce more heteropatriarchy. They suggest that a great deal of the work of formulating new political futures is in shifting cultural norms that have entrenched what they call the "cis-hetero-topia of post colonial Africa" (FAKA 2017). Siyakaka feminism is rooted in the celebration of nonnormativity and in the notion that freedom post-apartheid means the ability to own your body, particularly if you are femme or queer. In their own words, "Siyakaka is our manifesto for everything we do. Siyakaka means there's no apology. We will do what we want. Nobody will police our bodies" ("Desire Marea: Siyakaka"). In articulating these desires outside of normativity, FAKA embodies and performs a kwaito futurity. This is a future that is linked with the pussy parties, Babes Wodumo, Lebo Mathosa, Boom Shaka, Mandoza, Mafikizolo, izikhothane, my friends partying through Jozi, and audacious remembrances of the past. This is a future created through reconfiguring the past and invigorating the present. This is a future heavily invested not only in contesting what FAKA calls cis-hetero-topias, but also in imagining future queer femmetopias.

Freedom Remastered: On the Dismantling of Apartheid and the Afterlife of Kwaito

South Africa entered its transition to freedom aware of the failures of most postcolonial political economies to create sustainable futures for their citizenry. Part of the post-apartheid promise was the informed intention to do freedom differently. One central aspect of the post-apartheid liberation movement was the freeing of the body, particularly for those lacking class privilege, the youth, women, and queers. Over two decades later, it is clear that the project of South Africa is fraying. The post-apartheid compromise whereby the political economy remains unchanged and a Black elite and middle class are nurtured is collapsing under the entrenched hierarchies of race and class as well as under the weight of heteropatriarchy. What is unclear is what new political vernaculars will replace the current unworkable ones.

Kwaito enters the post-apartheid future as a cultural movement that on some fundamental level expressed an exhaustion with particular truisms of the relationship between cultural practices and politics. It reshapes this relationship in a number of different ways that I have explored in this text. Most importantly, kwaito asked post-apartheid South Africa about the sound and performance of freedom. What do South Africans get to do with that freedom? Can they dance differently to it? Kwaito asked the political establishment of South Africa if their conceptualization of freedom could accommodate everyone. And if it could not, then what kind of freedom was it anyway? In her text *Reflecting Rogue*, Pumla Gqola (2016) borrows the concept "normalizing freedom" from writer and literary critic Njabulo Ndebele. She suggests that part of the challenge of post-apartheid South Africa is moving from the struggle for freedom to moving toward what freedom should mean. As I have suggested, kwaito is a test of what freedom could entail. The performances that I highlight in this text, ranging from Boom Shaka's "Nkosi Sikelel' iAfrika" to Mafikizolo's "Khona," were all various attempts to remaster the performance of freedom. They were all variously critiqued and celebrated because of a range of kwaito embodiments. Kwaito is an important part of an effort to expand norms around South African liberation without in turn reproducing old norms for freedoms.

I return repeatedly to embodiment because the critiques of kwaito were often about performance and what these performances signified. Young people were using their bodies to articulate their freedoms while being told consistently that they had no right to do so. As FAKA reminds us, the penalties for inappropriate forms of embodiment are disproportionately born by women and queers; however, my analysis of Mandoza demonstrates that even the cishetero body bears its scars. Steve Biko (2002) and Audre Lorde (1984) have reminded us over the years that freedom must not simply be about assimilating Blacks into dominant power structures. For Biko, this assimilation of Blacks into dominant white societal norms produces new "unfreedoms" that re-create settler colonialism in blackface. For Lorde, this imitation was exemplary of the continued reliance on heteropatriarchy as a cultural and governing trope. Lorde cautioned against both the use of the past in ways that were mythic and the creation of a liberation project built on simplistic forms of cultural unity. She argued, "We forget that the necessary ingredient needed to make the past work for the future is our energy in the present, metabolizing one into the other" (1984, 136). She was concerned that so much of liberatory discourse in Black communities never reached the Fanonian moment of turning attention to the oppressor but instead remained turned inward to the policing of the community within, reproducing the structures of white supremacy. "As Black people, if there is one thing we can learn from the 60s, it is how infinitely complex any move for liberation must be. For we must move against not only those forces which dehumanize us from the outside, but also against those oppressive values which we have been forced to take into ourselves" (1984, 135). Lorde ultimately called for a respect for differences within community as central to any liberation project.

To conclude, I revisit kwaito bodies and FAKA's concept of Siyakaka feminism. Much like those who fought against apartheid, kwaito adherents put their bodies on the line. They experiment with freedom and push the boundaries of what they could and could not do. They expose the hypocrisies of post-apartheid South Africa. They reveal that freedom is complex and uncomfortable, but also that it is incomplete. They ask for an embodied future, one that would allow the nation to be performed to a kwaito beat. This new free South Africa would center the pleasures of young people, of women, of queers and the open possibilities their freedoms represented for the nation. This would be a South Africa that could account for the tsotsi and that understood that the past should be engaged creatively in the present as a way toward something new. Kwaito bodies push for the freedoms they desired, despite the criticisms and dismissals they suffered. In the process they have created a rubric for new political vernaculars and new cultural movements that are the engines of kwaito futurity. In a moment where the election of Cyril Ramaphosa (who oversaw the Marikana mine workers' massacre) as president of South Africa suggests more of the same (neo)liberal formation of the South African political economy with its attendant cis-hetero-topia, we need the pussy parties, we need Babes Wodumo to dance for our lives, and we need FAKA to perform fabulous alternative futures. It will be these new cultural eruptions that spring from kwaito that will do the cultural and metaphorical labor of birthing new South African freedoms. Boom Shaka challenged us back in 1994, reminding us that "it's about time to listen." By listening, dancing, singing, and offering my body to kwaito, I learned my most cherished lessons about freedom in post-apartheid South Africa. It is my humble hope that the South African nation—inspired by the vibrant afterlife of kwaito—will pay attention to the crucial lessons necessary for its long, free future.

Notes

4. The Kwaito Feminine

1. Few, if any, criticisms of Mathosa's performances have originated from outside Black communities.
2. For more information about Brenda Fassie, see Gqola 2004; Madondo 2014.
3. Fassie's final hit, "Yizo Yizo," speaks plaintively of the narrator's search for material and emotional comfort that seemed to be denied her.
4. Muñoz (1999) understood that disidentification was not always an adequate strategy of resistance or survival. There may be moments when minoritarian subjects need to be more openly resistant or the violence of the circumstances may require conformity.
5. One example of this ethics of knowledge is the teen game show *Jam Alley*, where contestants are rewarded for their musical knowledge and ability. Young people must demonstrate detailed knowledge about both South African and Afrodiasporic musicians. In addition they must competently perform songs in their given genre of choice (generally hip-hop, kwaito, and R&B) in order to be awarded the main prize. Several kwaito stars, most notably Mshoza, began their recording careers on this show.

6. Jackrolling, a term used to describe gang rape, particularly gang rape that involved abduction of the victim, became a media sensation in the mid- to late 1990s. Despite the newfound media attention on the topic, there was nothing new about either gang rape or the term "jackrolling" itself, which had been in circulation at least as far back as the 1980s. What was new was media attention to the sexual violation of Black women (by Black men) and the fact that with the dismantling of apartheid laws, these crimes, which were previously confined to township and rural areas, were now perpetrated in formerly all-white spaces such as university dormitories.
7. Consider this description of an early Boom Shaka performance: "One magazine (*Bona*, November 1996: 31–33) published an article probing whether such lewd routines were the result of taking illegal substances" (Stephens 2000, 269).
8. At one point, Jazz raps, "They [other suitors presumably] come rough and tough but it's never enough. You need a man in your life that's dangerous, adventurous."
9. It must be noted that the female initiation rites that Arnfred observed did not involve the cutting of any of the women's genitalia. However, part of the initiation rites did include the manipulation of female genitalia (in many instances, creating longer labia), a practice that Arnfred has stated enhances the sexual act for both men and women and the prohibition of which (according to UN standards on female genital mutilation) she finds baffling. It might also be equally interesting to note that the practice of vaginal rejuvenation, which in some cases involves the cutting and resculpting of the female genitalia in Western nations, has rarely been attacked as a form of female genital mutilation.
10. There is little evidence that female-female sensuality has the same kind of public meanings (at least among Black South Africans) that mark its incursion into fetishized space in Western sexual discourse, with the same-sex female interaction being performed for the presumed delight of the male spectator.
11. "Double adaptor" is a popular S'camto term for bisexual.
12. The Forum for the Equality of Women is a Black lesbian group addressing issues of violence as they relate predominantly to working-class, Black lesbian women. Many Black lesbians have been subject to rapes, beatings, and, in a recent case in Cape Town, death, as a result of publicly living their lives as lesbians.

5. The Black Masculine in Kwaito

1. Mandoza also noted that he did not perform hip-hop because of his lack of English fluency. Mandoza entered the music industry at a time when most hip-hop artists rapped in English or Afrikaans.
2. While not examined here, these forms of queer desire were also articulated to forms of female masculinities in South Africa. See Swarr 2012.

6. Mafikizolo and Youth Day Parties

1. Also important to think about in diasporic longing is the type of displacement that occurs as a result of migrant labor that was central to the colonial-apartheid project

of destroying independent African economies so that Black South Africans could become reserve labor for South African capital.

2. There seems to be some disagreement over whether three or four buildings survived the demolition. The Trevor Huddleston Memorial Centre cites four buildings as having survived even though most journalistic and scholarly accounts typically cite three surviving buildings. The tour of Sophiatown, sponsored by the Sophiatown Heritage and Cultural Centre, identifies three surviving buildings: Christ the King Anglican Church, Dr. A. B. Xuma's home, and a second home known now as the "ghost house." Also remaining is one wall of the old Odin Cinema, which the Huddleston Centre may be counting as a fourth remaining building.

3. Of course, it is unlikely, given the destruction of Sophiatown, that Matsho can actually go to his parents' old home (unless his parents' old home is the "ghost house") and attempt to reclaim it. Rather, his quote speaks to the sense of loss and how memory operates in claims over space.

4. Judging by the airplay on South African radio, R&B is, however, a popular genre of music in South Africa. There just have been few South African artists who have been able to successfully record and sell albums in the genre and its popularity is often limited to American (and to a lesser extent British) singers.

5. While it is not credited, it appears that the video takes place in Lesedi Cultural Village, a lodge and tourist site that features a number of replica "traditional" villages representing a cross-section of South Africa's ethnolinguistic groups.

6. Marabi music, developed in the 1920s and 1930s, was heavily influenced by American ragtime and blues but also drew heavily on local sonic traditions. Marabi typically referred to the style of keyboard playing that was central to the sonic texture of the music. It relied on repetitive extended chord progressions (a common feature of traditional folk music) in order to create harmonies that would be easy for crowds in shebeens to pick up and dance to. Marabi laid the foundation for later South African jazz music. For more on Marabi music and its musical, social, and political influence, see Ballantine 1993.

7. Kofifi was another name for Sophiatown. Contemporary (2014) manifestations of Kofifi parties abound, with Kofifi Entertainment sponsoring one of the more popular ones. The organizers describe their events as "old school vintage themed parties with a combination of Sophia Town, Pantsula, and Tap Dance Musical Experience with the Hottest DJs." For more information, see https://www.facebook.com/KofifiMovement.

8. While perhaps made famous by Madonna, vogueing is a queer African American art form. For more on vogueing and queer African American performance, see Bailey 2013. Waacking is also an African American dance form that developed out of the disco scene in 1970s Los Angeles and later influenced early hip-hop dancing. While not explicitly queer, it had queer components. You can see evidence of early waacking featured prominently in old *Soul Train* videos available online. For an example, see "Soul Train Shabba Doo Dance Routine," https://www.youtube.com/watch?v=B-WG0gdg2QA (accessed September 17, 2019).

9. I am speaking here particularly of the number of debates within the Anglophone Caribbean, and Western centers of Afrodiasporic migration (Toronto, New York,

London, Paris), and African countries such as Nigeria and Uganda that have politicized queer sexualities and attempted to legislate them out of existence
10. "Official Mafikizolo ft. Uhuru Khona."
11. "Official Mafikizolo ft. Uhuru Khona."
12. Youth Day is the national holiday approved by the post-apartheid ANC government to celebrate youth and to commemorate the uprising of June 16.
13. Zuma was accused of rape by Fezekile Kuzwayo, the daughter of a deceased ANC comrade who regarded Zuma as a father figure. Surely, she must have assumed that his home was a safe space for her. Zuma was acquitted of the crime and the victim is now deceased, but her memory is often invoked by feminist organizers in South Africa as an example of the normalization of rape and violence against women in South Africa.

Glossary

AMA KIP KIP: Candied popcorn that comes in a variety of colors and is sold in small plastic bags. Often considered a treat for children. Typically sold in local convenience shops (tuck shops) common in working-class Black residential areas. It is also the name of a South African–based streetwear fashion line.

ANC: African National Congress. The party that has ruled South Africa since its transition from apartheid.

BEE: Black Economic Empowerment. A program developed by the post-apartheid government to address economic inequality. Rather than challenging neoliberal capitalism, it seeks to diversify those who might benefit from the economy. In theory, companies are required to diversify management and ownership opportunities to communities disadvantaged by apartheid (predominantly on the basis of race). Critics claim that the program has resulted in very little shift in the ownership of South Africa's economy; instead, a small elite has been enriched at the expense of the masses.

JOHNNY CLEGG (1953–2019): A white musician who gained fame for his ability to play in Black musical and performance idioms. During apartheid, his insistence on multiracial bands and performance venues made him a target of the state. His engagement with Black South Africans (proficiency in African languages, music, and dance) is often cited as an example of an idealized white South African performance of identity.

DA: Democratic Alliance. South Africa's official opposition party. It has gained increasing influence as the ANC's grip on power has weakened as a consequence of the numerous political scandals that have surrounded the Zuma administration. The DA traces its origins to white liberal movements that opposed apartheid in the 1950s. While ostensibly multiracial, the DA struggles to shed the image that it operates in the interests of wealthy white South Africans and other non-Black African racial groups (Indians, Coloureds).

EKASI: Township; literally "the place of the township."

GINI COEFFICIENT: An economic formula used to measure income and/or wealth distribution in a given country; often serves as a proxy to determine economic inequality.

G'QOM: A music form that developed out of Durban post-2000 and that has recently gained more prominence in South African popular music. Like kwaito, it is considered to be a subgenre within the larger international house music idiom. The beats are sparse and broken down, providing a strong backing track for dancing and improvisational wordplay.

HOLA 7: A slang greeting/farewell popular in Johannesburg and surrounding areas during the early to mid-2000s. The actor and kwaito star Zola played with the term as the name of his show, which he called *Zola 7*.

KLEVA: A South African slang term (most likely derived from the English word "clever") for a streetwise young man.

KOFIFI: A slang (Tsotsitaal) word used to describe Sophiatown.

KOTA: A hollowed-out quarter loaf of bread filled with French fries, cheese, lunchmeat (usually bologna), and atchar. It is a popular township fast food.

LOBOLA: Often translated as bride-price; the amount of money paid by the groom's family to the bride's family as a symbolic gesture of recognition and good faith in the union of two families. Though it is traditionally paid in cows, money, household goods, and consumer items now can be used to satisfy lobola payments.

MAJITA: From Tsotsitaal/S'camto, meaning "guys."

DOROTHY MASUKA (1935–2019): South African jazz singer who was extremely popular in the 1950s Sophiatown scene. She is seen as one of the pioneers of Black women's singing performance in South African popular music. The apartheid government saw her as a threat and banned her songs. She went into exile for three decades. She is now remembered as a key figure of resistance and women's empowerment.

NORWOOD: A northern suburb neighborhood of Johannesburg about seven kilometers northeast of the Central Business District. In the post-apartheid period, as neighboring Orange Grove became increasingly economically, racially, and internationally diverse, Norwood retained its upscale image and its reputation for high-end restaurants. In the early 2000s, it became a site for contested ideas of public space usage regarding nightlife, hosting a youthful multiracial and increasingly Black crowd.

PEOPLE OF THE SOUTH: Hosted by Dali Tambo (son of the prominent ANC leader Oliver Tambo), the show aired from 1994 and 2002 (and again from 2012 to 2013). It featured leading figures in South African arts and politics and made a point to have a multiracial set of guests, in many ways creating the idea of a new multiracial South African public culture. Many prominent figures in entertainment, including several kwaito stars, were guests on the show in the late 1990s and early 2000s.

RAND SHOW: The largest consumer exhibition show in South Africa. Begun in 1894, it is typically held around the Easter holidays in the NASREC (National Recreational) center, located between the southern suburban neighborhoods of Johannesburg and Soweto. A number of events associated with the show, including a fair and concerts, often cater to a different crowd than the consumer exhibition itself.

SAMA: South African Music Awards. Similar to the Grammy Awards, the SAMAs recognize top accomplishments in the field of music.

S'CAMTO: A slang language deriving much of its vocabulary from local languages, usually either a Nguni- (Zulu-Xhosa) or Sotho- (Sesotho-Setswana) based slang. It is distinctive from Tsotsitaal in that it derives many of its words and much of its logic from African languages rather than Afrikaans. Considered a cant language by many linguistics, its vocabulary and usage often exclude those (elders, those from rural areas, women) who are not adept in speaking it.

SHARPEVILLE MASSACRE: On March 21, 1960, apartheid authorities fired on peaceful protesters who had organized against the Pass Laws, which sought to control the free movement of Black labor, killing sixty people. The apartheid government subsequently cracked down on anti-apartheid protest, banning most Black and multiracial political organizations and the Communist Party.

SHEBEEN: Originally informal taverns often located in homes and run by Black women who brewed liquor traditionally. Colonial and apartheid authorities banned these businesses in an attempt to monopolize the alcohol trade. Today, "shebeen" tends to be used more generally to refer to smaller-scale formal and informal establishments that sell liquor in Black areas.

TSOTSITAAL: Literally "tsotsi language"; an Afrikaans-based slang that originated in a number of nonstandard versions of Afrikaans mixed with local African languages.

TUCK SHOP: A local convenience store typically operated out of a home or small building.

YEOVILLE: A bohemian neighborhood located just beyond the Johannesburg's Central Business District (three kilometers north of downtown) with a history of serving as an incubator of countercultural artistic and social movements. Along with Berea and Hillbrow, Yeoville became one of the first neighborhoods to open up to mixed-race living when residential segregation laws began to be ignored and relaxed. During the 1990s, it was one of the centers of the birth of South Africa's new youth cultures, including kwaito.

References

Ahmed, N. 2004. "It's Taxi War on Miniskirts." *Daily Sun* (October 4): 1–2.
Akoonyatse, Bakang. 2016. "Life Lessons from Africa's Original Bad Girls." *True Africa* (January 21). https://trueafrica.co/article/life-lessons-africas-original-bad-girls/.
Allen, Jafari Sinclaire. 2011. *Venceremos? The Erotics of Black Self-Making in Cuba*. Durham, NC: Duke University Press.
Allen, Lara. 2004. "Kwaito versus Crossed-Over: Music and Identity during South Africa's Rainbow Years, 1994–1999." *Social Dynamics: A Journal of African Studies* 30 (2): 82–111.
Appadurai, Arjun. 1996. *Modernity at Large: Cultural Dimensions of Globalization*. Minneapolis: University of Minnesota Press.
Arnfred, Signe. 2003. "Contested Constructions of Female Sexualities: Meanings and Interpretations of Initiation Rituals." International Association for the Study of Sexuality, Culture and Society international conference. University of Witwatersrand, Johannesburg.
Ashforth, Adam. 1999. "Weighing Manhood in Soweto." CODESRIA *Bulletin* (3–4): 51–58. http://www.ajol.info/index.php/codbull/issue/view/2922.

Bailey, Marlon M. 2005. "The Labor of Diaspora: Ballroom Culture and the Making of a Black Queer Community." PhD dissertation, University of California, Berkeley.

Bailey, Marlon M. 2013. *Butch Queens Up in Pumps: Gender, Performance, and Ballroom Culture in Detroit.* Ann Arbor: University of Michigan Press.

Ballantine, Christopher. 1993. *Marabi Nights: Early South African Jazz and Vaudeville.* Johannesburg: Ravan Press.

Ballantine, Christopher. 1999. "Looking to the USA: The Politics of Male Close-Harmony Song Style in South Africa during the 1940s and 1950s." *Popular Music* 12:1–17.

Barnett, Clive. 2004. "Yizo Yizo: Citizenship, Commodification and Popular Culture in South Africa." *Media, Culture and Society* 26 (2): 251–271.

Biko, Steve. 2002. *I Write What I Like: Selected Writings.* Chicago: University of Chicago Press.

Blignaut, Charl. 1998. "Boom Shaka Shake It Up." *Mail and Guardian* (May 15). www.mg.co.za/article/1998-05-15-boom-shaka-shake-it-up.

Blignaut, Charl. 2016. "Lebo Mathosa: Freedom's Poster Girl." *City Press* (October 23). www.channel24.co.za/Music/News/lebo-mathosa-freedoms-poster-girl-20161023.

Blose, Maud. 2012. "Pornographic Objectification of Women through Kwaito Lyrics." *Agenda* 26 (3): 50–60.

Bogatsu, Afrika. 2016. "At Pussy Party, There's No Room to Be a Dick." *OkayAfrica* (October 19). http://www.okayafrica.com/pussy-party-johannesburg-south-africa/.

Boikanyo, Refilwe. 2013. "We Ask: Can *Ikhothane* Bragging Battles Become Performance Art?" *Elle Magazine* [South Africa] (February): 25–29.

Bond, Patrick. 2000. *Cities of Gold, Townships of Coal: Essays on South Africa's New Urban Crisis.* Trenton, NJ: Africa World Press.

Bond, Patrick, and Shauna Mottiar. 2013. "Movements, Protest, and a Massacre in South Africa." *Journal of Contemporary African Studies* 31 (2): 283–302.

Botha, Andrea. 2006. "Mandoza 'Too Expensive'—Hood." *News 24* (March 17). http://www.news24.com/SouthAfrica/News/Mandoza-too-expensive-Hood-20060317.

Bourdieu, Pierre. 1986. "The Forms of Capital." In *Handbook of Theory and Research for the Sociology of Education*, edited by John G. Richardson, 241–258. New York: Greenwood.

Bowes, Greg. 2015. "Born Free, South Africa and the House Nation." *Boiler Room* (February 27). https://boilerroom.tv/born-free-south-africa-the-house-nation/.

Brah, Avtar. 1996. *Cartographies of Diaspora: Contesting Identities.* New York: Routledge.

Brookes, Heather J. 2001. "O Clever 'He's Streetwise.' When Gestures Become Quotable: The Case of the Clever Gesture." *Gesture* 1 (2): 167–184.

Brown, D. 2003. "Lil' Kim Interview." *Rapreviews.com* (May 13). www.rapreviews.com/interview/kim.html.

Brown Shuga. 2011. "At the SAMA Nominees Party." http://www.justcurious.co.za/2011/04/at-the-sama-nominees-party/. Accessed February 5, 2014.

Browne, Simone. 2015. *Dark Matters: On the Surveillance of Blackness.* Durham, NC: Duke University Press.

Bushman, Casey. 2012. "Elusive Urbanity: Shopping for Global Prominence, Personal Liberation, and Public Space in Soweto." Master's thesis, Indiana University.

Butler, Judith. 1993. *Bodies That Matter: On the Discursive Limits of "Sex."* New York: Routledge.

Central Intelligence Agency (CIA). 2013. *The World Factbook.* www.cia.gov/library/publications/the-world-factbook/rankorder/2172rank.html.

Chibba, Reesha. 2006. "Telling the Story of Sophiatown." *Mail and Guardian* (February 11). http://mg.co.za/article/2006-02-11-telling-the-story-of-sophiatown.

Collins, Patricia Hill. 1990. *Black Feminist Thought: Knowledge, Consciousness, and the Politics of Empowerment.* New York: Routledge.

Conquergood, Dwight. 2002. "Lethal Theatre: Performance, Punishment, and the Death Penalty." *Theatre Journal* 54 (3): 339–367.

Conquergood, Dwight. 2006. "Rethinking Ethnography: Towards a Critical Cultural Politics." In *The Sage Handbook of Performance Studies*, edited by D. S. Madison and J. Hamera, 351–365. Thousand Oaks, CA: Sage.

Cooper, Carolyn. 2004. *Sound Clash: Jamaican Dancehall Culture at Large.* New York: Palgrave Macmillan.

Coplan, David. 1985. *In Township Tonight: South Africa's Black City Music and Theater.* New York: Longman Press.

Coplan, David. 2000. "Popular History; Cultural Memory." *Critical Arts: South-North Cultural and Media Studies* 14 (2): 122–144.

Coplan, David. 2005. "God Rock Africa: Thoughts on Politics in Popular Black Performance in South Africa." *African Studies* 64 (1): 9–27.

Coplan, David. 2008. "Popular Music in South Africa." In *The Garland Handbook of African Music*, edited by Ruth M. Stone, 406–428. New York: Routledge.

Cox, Aimee Meredith. 2015. *Shapeshifters: Black Girls and the Choreography of Citizenship.* Durham, NC: Duke University Press.

Crenshaw, Kimberlé. 1989. "Demarginalizing the Intersection of Race and Sex: A Black Feminist Critique of Antidiscrimination Doctrine, Feminist Theory, and Antiracist Politics." *University of Chicago Legal Forum* 1989 (1): 139–167.

Da Silva, Issa Sikiti. 2007. "Maponya Mega-Mall Opens in Soweto." *BizCommunity* (September 28). www.bizcommunity.com/Article/196/160/18424.html.

Davie, Lucille. 2006. "Sophiatown Again, 50 Years On." February 14. https://web.archive.org/web/20081207011053/http://www.southafrica.info/about/history/sophiatown140206.htm.

Delaney, Samuel. 1999. *Times Square Red, Times Square Blue.* New York: New York University Press.

Desai, Ashwin. 2002. *We Are the Poors: Community Struggles in Post-Apartheid South Africa.* New York: New York University Press.

Diawara, Manthia. 1998. *In Search of Africa.* Cambridge, MA: Harvard University Press.

Diawara, Manthia. 2003. *We Won't Budge: An African Exile in the World.* New York: Basic Civitas Books.

Dlamini, Jacob. 2009. *Native Nostalgia.* Johannesburg: Jacana Media.

Dolby, Nadine. 2006. "Popular Culture and Public Space in Africa: The Possibilities of Cultural Citizenship." *African Studies Review* 49 (3): 31–47.

"Don't Drink on Youth Day." 2003. *News 24* (June 11). www.news24.com/SouthAfrica/News/Dont-drink-on-Youth-Day-20030610.

Dube, Pamela. 2001. "Winnie Comes Out Fighting." *IOL* (June 23). http://www.iol.co.za/news/politics/winnie-comes-out-fighting-1.68564?ot=inmsa.ArticlePrintPageLayout.ot.

Du Bois, W. E. B. 2001. *The Negro*. Philadelphia: University of Pennsylvania Press.

The Edge. 2016. "Why Babes Wodumo Calls Herself 'Ratchet.'" *The Edge Search* (September 15). https://www.theedgesearch.com/2016/09/why-babes-wodumo-calls-herself-ratchet.html.

Edwards, Brent Hayes. 2003. *The Practice of Diaspora: Literature, Translation, and the Rise of Black Internationalism*. Cambridge, MA: Harvard University Press.

Ellapen, Jordache. 2007. "The Cinematic Township: Cinematic Representations of 'Township Space' and Who Can Claim the Rights to Representation in Post-Apartheid South African Cinema." *Journal of African Cultural Studies* 19 (1): 113–138.

Erasmus, Zimitri, ed. 2001. *Coloured by History, Shaped by Place: New Perspectives on Coloured Identities in Cape Town*. Cape Town: Kwela.

Erlmann, Veit. 1999. *Music, Modernity and the Global Imagination: South Africa and the West*. Oxford: Oxford University Press.

Eyerman, Ron. 2001. *Cultural Trauma: Slavery and the Formation of African American Identity*. Cambridge: Cambridge University Press.

Eyerman, Ron, and Andrew Jamison. 1998. *Music and Social Movements: Mobilizing Traditions in the Twentieth Century*. Cambridge: Cambridge University Press.

Fabian, Johannes. 2002. *Time and the Other: How Anthropology Makes Its Object*. New York: Columbia University Press.

FAKA. n.d. "About." *Siyakaka.com*. www.siyakaka.com/about-1. Accessed August 29, 2019.

Fanon, Frantz. 1963. *The Wretched of the Earth*. New York: Grove Press.

Farber, Leora. 2010. "Africanising Hybridity? Toward an Afropolitan Aesthetic in Contemporary South African Fashion Design." *Critical Arts* 24 (1): 128–167.

Fikentscher, Kai. 2000. *You Better Work! Underground Music in New York City*. Middletown, CT: Wesleyan University Press.

Fink, Katharina. 2014. "Remix Sophiatown: Re-Imagining Can Themba's The Suit in Contemporary South African Literature." In *Hospitality and Hostility in the Multilingual Global Village*, edited by Kathleen Thorpe et al., 285–303. Stellenbosch, South Africa: African Sun Media.

Foucault, Michel. 1987. "The Ethic of Care for the Self as a Practice of Freedom: An Interview with Michel Foucault." *Philosophy and Social Criticism* 12 (2–3): 112–131.

Fourie, Salomie. 2005. "South African Streetwear." *iFashion* (April 1). http://www.ifashion.co.za/index.php?option=com_content&task=view&id=461&Itemid=43.

The Fuss. 2016. "Rosie Parade and the Power of a Smile." *The Fuss* (May 30). http://thefuss.co.za/rosie-parade-everyday-hero/.

Gershon, Tanja. 2015. "To the Street: Passbooks, Permits and the Art of Public Life." A-R-P-A Journal 3 (July 3). http://www.arpajournal.net/to-the-street/.

Gilbert, Jeremy, and Ewan Pearson. 1999. *Discographies: Dance Music Culture and the Politics of Sound*. New York: Routledge.

Gilman, Sander. 1985. "Black Bodies, White Bodies: Towards an Iconography of Female Sexuality in the Late Nineteenth-Century Art, Medicine, and Literature." *Critical Inquiry* 12:204–242.

Gilroy, Paul. 1993. *The Black Atlantic: Modernity and Double Consciousness*. Cambridge, MA: Harvard University Press.

Gilroy, Paul. 2004. *Postcolonial Melancholia*. New York: Columbia University Press.

Godinho, Thorne. 2013. "Khona: Challenging Masculinity in Africa." *Vada* (September 4). http://vadamagazine.com/04/09/2013/television/khona.

Gqola, Pumla Dineo. 2004. "When a Good Black Woman Is Your Weekend Special: Brenda Fassie, Sexuality and Performance." In *Under Construction: "Race" and Identity in South Africa Today*, 139–148. Sandton, South Africa: Heinemann.

Gqola, Pumla Dineo. 2007. "How the 'Cult of Femininity' and Violent Masculinities Support Endemic Gender Based Violence in Contemporary South Africa." *African Identities* 5 (1): 111–124.

Gqola, Pumla Dineo. 2015. *Rape: A South African Nightmare*. Johannesburg: Jacana.

Gqola, Pumla Dineo. 2016. *Reflecting Rogue: Inside the Mind of a Feminist*. Johannesburg: Jacana.

Gregg, Robert. 2001. "Afterword." In W. E. B. Du Bois, *The Negro*, 245–272. Philadelphia: University of Pennsylvania Press.

Guilbault, Jocelyne. 2005. "Audible Entanglements: Nation and Diasporas in Trinidad's Calypso Music Scene." *Small Axe* 9 (1): 40–63.

Guma, Lance. 2012. "Zimbabwe: Detainees Allege Four Deaths at Lindela Centre." *SW Radio Africa* (August 21). https://allafrica.com/stories/201208220301.html.

Habib, Adam. 1997. "South Africa—The Rainbow Nation and the Prospects for Consolidating Democracy." *African Journal of Political Science* 2 (2): 15–37.

Hall, Stuart. 1993. "What Is This Black in Black Popular Culture?" *Social Justice* 20 (1–2): 104–114.

Hamm, Charles. 1988. *Afro American Music, South Africa and Apartheid*. ISAM Monographs No. 28. Brooklyn: Conservatory of Music, Brooklyn College of the City University of New York.

Hannerz, Ulf. 1994. "Sophiatown the View from Afar." *Journal of Southern African Studies* 20:181–193.

Hans, Bongani. 2013. "Zuma: Drugs, Booze Now the Enemy." *IOL* (June 17). http://www.iol.co.za/news/politics/zuma-drugs-booze-now-the-enemy-1.1533290#.VLShRYrF_vQ.

Hansen, Thomas Blom. 2006. "Sounds of Freedom: Music, Taxis, and Racial Imagination in Urban South Africa." *Public Culture* 18 (1): 185–208.

Harmse, Liana. 2014. "South Africa's Gini Coefficient: Causes, Consequences and Possible Responses." PhD dissertation, University of Pretoria.

Hart, Gillian. 2002. *Disabling Globalization: Places and Power in Post-Apartheid South Africa*. Berkeley: University of California Press.

Haupt, Adam. 2012. *Static: Race and Representation in Post-Apartheid Music, Media and Film*. Pretoria: HSRC Press.

Holsey, Bayo. 2008. *Routes of Remembrance: Refashioning the Slave Trade in Ghana*. Chicago: University of Chicago Press.

hooks, bell. 1996. *Reel to Real: Race, Sex, and Class at the Movies*. New York: Routledge.

Hope, Donna. 2006. "Dons and Shottas: Performing Violent Masculinity in Dancehall Culture." *Social and Economic Studies* 55 (1/2): 115–131.

Impey, Angela. 2001. "Resurrecting the Flesh? Reflections on Women in Kwaito." *Agenda* 16 (49): 44–50.

Institute for Justice and Reconciliation. 2012. *Transformation Audit 2012*. Cape Town: IJR.

Iqani, Mehita. 2015. "Agency and Affordability: Being Black Middle Class in South Africa in 1989." *Critical Arts: South-North Cultural and Media Studies* 29 (2): 126–145.

Iton, Richard. 2008. *In Search of the Black Fantastic: Politics and Popular Culture in the Post–Civil Rights Era*. Oxford: Oxford University Press.

Jaji, Tsitsi. 2014. *Africa in Stereo: Modernism, Music, and Pan-African Solidarity*. Oxford: Oxford University Press.

Johnson, E. Patrick. 2001. "'Quare' Studies, or (Almost) Everything I Know about Queer Studies I Learned from My Grandmother." *Text and Performance Quarterly* 21 (1): 1–25.

Johnson, E. Patrick. 2003. *Appropriating Blackness: Performance and the Politics of Authenticity*. Durham, NC: Duke University Press.

Johnson, E. Patrick. 2006. "Black Performance Studies: Genealogies, Politics, Futures." In *The Sage Handbook of Performance Studies*, edited by D. S. Madison and J. Hamera, 446–463. Thousand Oaks, CA: Sage.

Jones, Joni L. 1996. "The Self as Other: Creating the Role of Joni the Ethnographer for Broken Circles." *Text and Performance Quarterly* 16 (2): 131–145.

Jones, Joni L. 2002. "Performance Ethnography: The Role of Embodiment in Cultural Authenticity." *Theatre Topics* 12 (1): 1–15.

Jones, Omi Osun Joni L. 2006. "Performance and Ethnography, Performing Ethnography, Performance Ethnography." In *The Sage Handbook of Performance Studies*, edited by D. S. Madison and J. Hamera, 339–345. Thousand Oaks, CA: Sage.

Jones, Omi Osun Joni L. 2016. "The Role of Performance Studies in Black Studies, or What Do Bodies Have to Do with Intellectual Inquiry?" Paper presented at the University of Texas Annual Africa Conference, Austin, April.

Jules-Rossette, Bennetta, and David Coplan. 2004. "'Nkosi Sikelel' iAfrika': From Independent Spirit to Political Mobilization." *Cahier d'Etudes Africaines* 44 (173–174): 343–367.

Kabir, Ananya. 2004. "Musical Recall: Postmemory and the Punjabi Diaspora." *Alif: Journal of Comparative Poetics* 24:172–189.

Kaganof, Aryan. 2006a. "The Kwaito Story: Mafikizolo Interviewed by Aryan Kaganof." web.archive.org/web/20071103091146/http://kaganof.com/kagablog/2006/09/18/the-kwaito-story-mafikizolo-interviewed-by-aryan-kaganof/

Kaganof, Aryan. 2006b. "The Kwaito Story: Mandoza Interviewed by Aryan Kaganof." web.archive.org/web/20080223081445/http://kaganof.com/kagablog/2006/10/15/the-kwaito-story-mandoza-interviewed-by-aryan-kaganof/.

KasieKulture. 2008. "Vagina Monologues." *Kasiekulture!* (October 7). http://kasiekulture.blogspot.com/2008/10/feature.html.

Kelley, Robin D. G. 1997. *Yo' Mama's Disfunktional! Fighting the Culture Wars in Urban America.* Boston: Beacon.

Kempadoo, Kemala. 2004. *Sexing the Caribbean: Gender, Race, and Sexual Labor.* New York: Routledge.

Kharsany, Safeeyah. 2013. "Joblessness Plagues South Africa's Youth." *Al Jazeera* (June 18). http://www.aljazeera.com/indepth/features/2013/06/201361875748330888.html.

Kruger, Loren. 1995. "The Uses of Nostalgia: Drama, History and Liminal Moments in South Africa." *Modern Drama* 38:60–70.

Kunnie, Julian. 2000. *Is Apartheid Really Dead? Pan Africanist Working Class Cultural Critical Perspectives.* Boulder, CO: Westview.

Laccino, Ludovica. 2016. "Jacob Zuma Leads Tributes to South African Kwaito Legend Mandoza Who Has Died at 38." *International Business Times* (September 19). https://www.ibtimes.co.uk/jacob-zuma-leads-tributes-south-african-kwaito-legend-mandoza-who-has-died-38-1582057.

Lefebvre, Henri. 1996. *Writings on Cities.* Cambridge, MA: Blackwell.

Leiman, Layla. 2015. "Young South Africa: FAKA Redefining Representations of Black Queer Identity. *Between 10 and 5* (June 29). https://10and5.com/2015/06/29/young-south-africa-faka-your-representation-of-black-queer-identity.

Lindwa, Buhle. 2016. "Babes Wodumo Is the Original Carefree Black Girl. Give Her the Credit She Deserves." *LiveMagSA* (August 4). https://livemag.co.za/babes-wodumo-original-carefree-black-girl/.

Lipsitz, George. 1994. *Dangerous Crossroads: Popular Music, Postmodernism, and the Poetics of Place.* New York: Verso.

Livermon, Xavier. 2012. "Queer(y)ing Freedom: Black Queer Visibilities in Postapartheid South Africa." *GLQ: A Journal of Lesbian and Gay Studies* 18 (3): 297–323.

Livermon, Xavier. 2014. "Soweto Nights: Making Black Queer Space in Post-Apartheid South Africa." *Gender, Place, and Culture: A Journal of Feminist Geography.* http://dx.doi.org/10.1080/0966369X.2013.786687.

Livermon, Xavier. 2015. "Usable Traditions: Creating Sexual Autonomy in Post-apartheid South Africa." *Feminist Studies* 41 (1): 14–41.

Lynnée Denise. 2012. "The Afro-Digital Migration: House Music in Post Apartheid South Africa." http://www.djlynneedenise.com/menu.

Lorde, Audre. 1984. *Sister Outsider: Essays and Speeches by Audre Lorde.* Berkeley, CA: Crossing Press.

Mabandu, Percy. 2012. "Brash Bling and Ghetto Fabulous." *City Press* (October 6). https://www.news24.com/Archives/City-Press/Brash-bling-and-ghetto-fabulous-20150429.

Macharia, Keguro. 2016. "Political Vernaculars: Freedom and Love." *New Inquiry* (March 14). https://thenewinquiry.com/political-vernaculars-freedom-and-love/.

Madondo, Bongani. 1998. "Controversial Boom Shaka Gets the Whole Town Jiving." *City Press*. http://152.111.1.87/argief/berigte/citypress/1998/06/14/25/1.html. Accessed June 6, 2012.

Madondo, Bongani, ed. 2014. *I'm Not Your Weekend Special: Portraits on the Life + Style & Politics of Brenda Fassie*. Johannesburg: Picador Africa. Kindle edition.

Magubane, Zine. 2003. "The Influence of African American Cultural Practices in South Africa 1890–1990." In *Leisure in Urban Africa*, edited by Paul Zeleza and Cassandra Veney, 297–319. Trenton, NJ: Africa World Press.

Mahlerbe, Petrus. 2014. "Give Youth Day Back to the Kids." IOL (June 16). http://www.iol.co.za/pretoria-news/opinion/give-youth-day-back-to-the-kids-1.1704038#.VLSgoorF_vQ.

Mahmood, Saba. 2001. "Feminist Theory, Embodiment, and the Docile Agent: Some Reflections on the Egyptian Islamic Revival." *Cultural Anthropology* 16 (2): 202–236.

Mail and Guardian. 1998. "Sacrilege or Success? What People Think of Boom Shaka's Nkosi Sikelel' iAfrika." May 15. http://mg.co.za/print/1998-05-15-sacrilege-or-success.

Makgoba, Malegapuru. 2005. "Wrath of the Dethroned White Males." *Mail and Guardian* (March 25). http://mg.co.za/article/2005-03-25-wrath-of-dethroned-white-males.

Makhulu, Anne-Maria. 2015. *Making Freedom: Apartheid, Squatter Politics, and the Struggle for Home*. Durham, NC: Duke University Press.

Malatji, Ngwako. 2013. "Mandoza on the Skids." *City Press* (October 27). https://www.news24.com/Archives/City-Press/Mandoza-on-the-skids-20150429.

Manuel, Peter. 1998. "Gender Politics in Caribbean Popular Music: Consumer Perspectives and Academic Interpretation." *Popular Music in Society* 22 (2): 11–29.

Marumo, K. 2006. "Hip-Hop Beefcake." *Rage*. www.rage.co.za/issue43/2006/03hiphopbeefcake.html. Accessed July 23, 2006.

Matory, J. Lorand. 2005. *Black Atlantic Religion: Tradition, Transnationalism, and Matriarchy in the Afro-Brazilian Candomble*. Princeton, NJ: Princeton University Press.

Mattera, Don. 1987. *Gone with the Twilight: A Story of Sophiatown*. London: Zed.

Mazibuko, Lindiwe. 2011. "On Freedom through Opportunity." *Politicsweb* (November 29). https://www.politicsweb.co.za/politics/on-freedom-through-opportunity--lindiwe-mazibuko.

Mbembe, Achille. 2002. "African Modes of Self Writing." *Public Culture* 14 (1): 239–273.

Mbembe, Achille, Nsizwa Dlamini, and Grace Khunou. 2008. "Soweto Now." In *Johannesburg: The Elusive Metropolis*, edited by Sarah Nuttall and Achille Mbembe, 239–247. Durham, NC: Duke University Press.

McBride, Dwight. 1998. "Can the Queen Speak? Racial Essentialism, Sexuality and the Problem of Authenticity." *Callaloo* 21 (2): 363–379.

McClintock, Anne. 1995. *Imperial Leather: Race, Gender, and Sexuality in the Colonial Context*. New York: Routledge.

McCloy, Maria. 2006. "Just Another SAMA Night." *Mail and Guardian* (May 12). https://mg.co.za/article/2006-05-12-just-another-sama-night.

McCune, Jeffrey Q. 2014. *Sexual Discretion: Black Masculinity and the Politics of Passing*. Chicago: University of Chicago Press.

McEachern, Montinique Denice. 2017. "Respect My Ratchet: The Liberatory Consciousness of Ratchetness." *Departures in Critical Qualitative Research* 6 (3): 78–89.

Mchunu, Harry. 2008. "Perpetuating the Stigma." *Witness* (February 22). https://www.news24.com/Archives/Witness/Perpetuating-the-stigma-20150430.

McKittrick, Katherine. 2006. *Demonic Grounds: Black Women and the Cartographies of Struggle.* Minneapolis: University of Minnesota Press.

Mhlambi, Thokozani. 2004. "Kwaitofabulous: The Study of a South African Genre." *Journal of the Musical Arts in Africa* 1 (1): 116–127.

Miller, Monica L. 2009. *Slaves to Fashion: Black Dandyism and the Styling of Black Diasporic Identity.* Durham, NC: Duke University Press.

Miller-Young, Mireille. 2014. *A Taste for Brown Sugar: Black Women in Pornography.* Durham, NC: Duke University Press.

Mkize, Vuyo. 2012. "Lebo's Music Rights Sold as Moms Feud." *IOL* (October 23). http://www.iol.co.za/news/crime-courts/lebo-s-music-rights-sold-as-moms-feud-1.1409326.

Mnisi, Jabulani. 2015. "Burning to Consume? Izikhothane in Daveyton as Aspirational Consumers." *Communicatio* 41 (3): 340–353.

Molele, C. 2005. "Planet Lebo." *Sunday Times* (March 6): 7.

Molobye, Kamogelo. n.d. "In Pursuit of a Differentiated Township Masculinity." *Our Queer Stories.* https://ourqueerstories.com/in-pursuit-of-a-differentiated-black-township-masculinity/. Accessed September 17, 2019.

Moore, Sara. 2009. "Pantsula—Dance and a Way of Life." *Happy People Dance.* happypeopledance.com/?p=315. Accessed December 26, 2012.

Moten, Fred. 2003. *In the Break: The Aesthetics of the Black Radical Tradition.* Durham, NC: Duke University Press.

Moynihan, Daniel P. 1965. *The Negro Family: The Case for National Action.* Washington DC: Office of Policy Planning and Research, US Department of Labor.

Mphithi, Luyolo. 2017. *Twitter* (March 5). https://twitter.com/LuyoloMphithi/status/838642791622258688.

Muñoz, José E. 1999. *Disidentifications: Queers of Color and the Performance of Politics.* Minneapolis: University of Minnesota Press.

Music.org.za. n.d. "Mandoza (South Africa)." http://music.org.za/artist.asp?id=94.

Namaste, Ki. 2000. *Invisible Lives: The Erasure of Transsexual and Transgendered People.* Chicago: University of Chicago Press.

National Public Radio (NPR). 2006. "The Voice of South African Kwaito: Zola." *NPR Music* (March 7). https://www.npr.org/templates/story/story.php?storyId=5249664.

Neal, Mark Anthony. 2003. "Tupac's Book Shelf: 'All Eyez on Me': Tupac Shakur and the Search for the Modern Folk Hero. W. E. B. Du Bois Institute for Afro-American Research, Harvard University April 17, 2003." *Journal of Popular Music Studies* 15 (2): 208–212.

Ngcaweni, Wandile. 2016. "How My Taxi Ride Differed from Sisonke Msimang's Uber Experience." *Daily Vox* (June 17). https://www.thedailyvox.co.za/wandile-ngcaweni-taxi-ride-differed-sisonke-msimangs-uber-ride/.

Ngudle, Amanda. 2004. "Lebo Mathosa Speaks Out: Drugs, Lesbians, and Lies." *Drum* (September 23): 15–16.

Nguyen, Vinh-Kim. 2005. "Uses and Pleasures: Sexual Modernity, HIV/AIDS, and Confessional Technologies in a West African Metropolis." In *Sex in Development: Science, Sexuality, and Morality in Global Perspective*, edited by Vincanne Adams and Stacy Leigh Pigg, 245–267. Durham, NC: Duke University Press.

Niaah, Sonjah Stanley. 2008. "Performance Geographies from Slave Ship to Ghetto." *Space and Culture* 11 (4): 343–360.

Niaah, Sonjah Stanley. 2009. "Negotiating a Common Transnational Space: Mapping Performance in Jamaican Dancehall and South African Kwaito." *Cultural Studies* 23 (5–6): 756–774.

Niaah, Sonjah Stanley. 2010. *Dancehall: From Slave Ship to Ghetto*. Ottawa: University of Ottawa Press.

Nixon, Rob. 1994. *Homelands, Harlem and Hollywood: South African Culture and the World Beyond*. New York: Routledge.

Nkosi, Lindokuhle. 2011. "Burn Swag Burn." *Mahala* (April 8). http://www.mahala.co.za/culture/burn-swag-burn/.

Nkosi, Lindokuhle. 2014. "Freedom Marches to Kwaito's Drum." *Mail and Guardian* (April 25). https://mg.co.za/article/2014-04-24-freedom-marches-to-kwaitos-drum.

Noyes, John K. 1998. "S/M in SA: Sexual Violence, Simulated Sex and Psychoanalytic Theory." *American Imago* 55 (1): 135–153.

Nuttall, Sarah. 2003. "Self and Text in Y Magazine." *African Identities* 1 (2): 235–251.

Nuttall, Sarah. 2004. "Stylizing the Self: The Y Generation in Rosebank, Johannesburg." *Public Culture* 16 (3): 430–452.

Nxedlana, Jamal. 2012. "Izikhothane: Two Sides to the Story." *Tumblr*. https://xosouthafrica-blog.tumblr.com/post/23551881482/izikhothane-two-sides-to-the-story-ubkhothane. Accessed September 17, 2019.

Okumu, Phiona. 2013. "Why Mafikizolo's New Song 'Happiness' Is Important." *Afripop! Global African* Culture (October). http://afripopmag.com/2013/10/why-mafikizolos-new-song-happiness-is-important/.

Oswin, Natalie. 2005. "Researching 'Gay Cape Town,' Finding Value Added Queerness." *Social and Cultural Geography* 6 (4): 567–586.

O'Toole, Kit. 2009. "What Does Remastering Mean Anyway?" *Blogcritics* (September 8). http://blogcritics.org/what-does-remastering-mean-anyway/.

Owen, Therese. 2013. "Reunited Mafikizolo Are Back with a Bang." *IOL* (March 13). http://www.iol.co.za/tonight/music/reunited-mafikizolo-are-back-with-a-bang-1.1485734.

Patterson, Orlando. 1991. *Freedom in the Making of Western Culture*. New York: Basic Books.

Peterson, Bhekizizwe. 2003. "Kwaito, 'Dawgs' and the Antimonies of Hustling." *African Identities* 1 (2): 197–213.

Pierre, Jemima. 2008. "'I Like Your Colour': Skin Bleaching and Geographies of Race in Urban Ghana." *Feminist Review* 90: 9–29.

Pierre, Jemima. 2012. *The Predicament of Blackness: Postcolonial Ghana and the Politics of Race*. Chicago: University of Chicago Press.

Pietilä, Tuulukki. 2013. "Body Politic: The Emergence of a 'Kwaito Nation' in South Africa." *Popular Music and Society* 36 (2): 143–161.

Pillay, Suren. 2008. "Dangerous Ordinary Discourse: Preliminary Reflections on Xenophobia, Violence and the Public Sphere in South Africa." CODESRIA *12th General Assembly, Cameroon* 7–11: 1–18.

Pokwana, Vukile. 1998. "Mixed Reaction over Funky National Anthem." *City Press.* http://152.111.1.87/argief/berigte/citypress/1998/05/10/5/2.html. Accessed June 6, 2012.

Posel, Deborah. 2010. "Races to Consume: Revisiting South Africa's History of Race, Consumption and the Struggle for Freedom." *Ethnic and Racial Studies* 33 (2): 157–175.

Ramackers, Helene. 2005. "Mandoza: I Am a Family Man Now." *True Love* (June): 73–77.

Ratele, Kopano. 2003. "Re tla Dirang ka Selo Se Ba Re Ho ke Ghetto Fabulous? Academics on the Streets." *Agenda* 17 (57): 46–50.

Reddy, Gayatri. 2005. *With Respect to Sex: Negotiating Hijra Identity in South Asia.* Chicago: University of Chicago Press.

Redmond, Shana. 2013. *Anthem: Social Movements and the Sound of Solidarity in the African Diaspora.* New York: New York University Press.

Roberts, Adam. 2002. "From Jo'burg to Jozi." In *From Jo'burg to Jozi: Stories about Africa's Infamous City,* edited by Heidi Holland and Adam Roberts, 1–14. Johannesburg: Penguin.

Rogerson, Christian M. 2006. "Developing the Fashion Industry in Africa: The Case of Johannesburg." *Urban Forum* 17 (3): 215–240.

Samuelson, Meg. 2007. *Remembering the Nation, Dismembering Women? Stories of the South African Transition.* Pietermaritzburg: UKZN Press.

Samuelson, Meg. 2008. "The Urban Palimpsest: Re-Presenting Sophiatown." *Journal of Postcolonial Writing* 44 (1): 63–75.

Santos, Dominique. 2013. "All Mixed Up: Music and Intergenerational Experiences of Social Change in South Africa." PhD dissertation, Goldsmiths, University of London.

Sarko, Anita. 1999. "A Big Ol' Music Story Starring Lil' Kim." *Interview* (November): 122–126.

Scott, David. 1999. *Refashioning Futures: Criticism after Postcoloniality.* Princeton, NJ: Princeton University Press.

Seabi, Mokgadi. 2011. "Pantsula Dancers Hit the Big Time in New Beyoncé Video." *City Press* (May 21). https://www.news24.com/Archives/City-Press/Pantsula-dancers-hit-the-big-time-in-new-Beyonce-video-20150430.

Sedumedi, Itumeleng. 2006. "Decency and the Entertainment Industry." *Matumza Online* (March 17). web.archive.org/web/20060427021917/http://matumzaonline.blogspot.com.

Seekane, K. 2006. "Maf Motswako." www.rage.co.za/issue43/. Accessed July 23, 2006.

Seidman Gay. 1993. "No Freedom without Women: Mobilization and Gender in South Africa 1970–1992." *Signs* 18 (2): 291–320.

Sesali, B. 2013. "Let's Get Ratchet! Check Your Privilege at the Door." *Feministing* (March 28). http://feministing.com/2013/003/28/lets-get-ratchet-check-your-privilege-at-the-door/.

Shabazz, Rashad. 2009. "'So High You Can't Get over It, So Low You Can't Get under It': Carceral Spatiality and Black Masculinities in the United States and South Africa." *Souls: A Critical Journal of Black Politics, Culture, and Society* 11 (3): 276–294.

Simone, AbdouMaliq. 2009. *City Life from Jakarta to Dakar: Movements at the Crossroads*. New York: Routledge.

Singer, Linda. 1993. *Erotic Welfare: Sexual Theory and Politics in the Age of Epidemic*. New York: Routledge.

Smith, Andrea. 2006. "Heteropatriarchy and the Three Pillars of White Supremacy: Rethinking Women of Color Organizing." In *The Colour of Violence: The Incite! Anthology*, edited by Incite!, 66–73. Cambridge, MA: South End Press.

Smith, Christopher Holmes. 1997. "Method in the Madness: Exploring the Boundaries of Identity in Hip-Hop Performativity." *Social Identities* 3 (3): 345–374.

Smith, Danyell. 1996. "Ghetto Fabulous." *New Yorker* (April 29–May 6): 50.

Snorton, C. Riley. 2014. "On the Question of Who's Out in Hip Hop." *Souls: A Critical Journal of Black Politics, Culture, and Society* 16 (3–4): 283–302.

"South African Artist Sets Sights on Conquering East Africa." 2008. *Artmatters.info*, (September 22). https://artmatters.info/2008/09/south-african-artist-sets-her-eyes-on-east-africa.

Stallings, L. H. 2013. "Hip Hop and the Black Ratchet Imagination." *Palimpsest: A Journal on Women, Gender, and the Black International* 2 (2): 135–139.

Steingo, Gavin. 2005. "South African Music after Apartheid: Kwaito, the 'Party Politic' and the Appropriation of Gold as a Sign of Success." *Popular Music and Society* 28:333–357.

Steingo, Gavin. 2007. "The Politicization of Kwaito: From the 'Party Politic' to Party Politics." *Black Music Research Journal* 27 (1): 23–44.

Steingo, Gavin. 2008a. "Historicizing Kwaito." *African Music* 8 (2): 76–91.

Steingo, Gavin. 2008b. "Producing Kwaito: Nkosi Sikelel' iAfrika after Apartheid." *World of Music* 50 (2): 103–120.

Steingo, Gavin. 2016. *Kwaito's Promise: Music and the Aesthetics of Freedom in South Africa*. Chicago: University of Chicago Press.

Stephens, Simon. 2000. "Kwaito." In *Senses of Culture*, edited by S. Nuttal and C. Michael, 256–277. Cape Town: Oxford University Press.

Story, Kaila. 2017. "Fear of a Black Femme: The Existential Conundrum of Embodying a Black Femme Identity While Being a Professor of Black, Queer, and Feminist Studies." *Journal of Lesbian Studies* 21 (4): 407–419.

Sutton, Rebecca, et al. 2011. "Waiting in Liminal Space: Migrants' Queuing for Home Affairs in South Africa." *Anthropology Southern Africa* 34 (1–2): 30–37.

Swarr, Amanda Locke. 2012. "Paradoxes of Butchness: Lesbian Masculinities and Sexual Violence in Contemporary South Africa." *Signs* 37 (4): 961–986.

Swartz, Sharlene. 2008. "Is Kwaito South Africa's Hip Hop? Why the Answer Matters and Who It Matters To." *World of Music* 50 (2): 15–33.

Tabane, Rapule. 2005. "The Zola Hall of Fame." *Mail and Guardian* (August 1). http://mg.co.za/article/2005-08-01-the-zola-hall-of-fame.

Taylor, Diana. 2003. *The Archive and the Repertoire: Performing Cultural Memory in the Americas*. Durham, NC: Duke University Press.

Thepa, Madala. 2004. "Up Close and Personal with Lebo Mathosa." *Sowetan Sunday World* (December 5): 4.

Thomas, Anthony. 1995. "The House the Kids Built: The Gay Black Imprint on American Dance Music." In *Out in Culture: Gay, Lesbian, and Queer Essays on Popular Culture*, edited by Corey K. Creekmur and Alexander Doty, 437–446. Durham, NC: Duke University Press.

Thomas, Deborah A. 2004. *Modern Blackness: Nationalism, Globalization, and the Politics of Culture in Jamaica*. Durham, NC: Duke University Press.

Titlestad, Michael. 2004. *Making the Changes: Jazz in South African Literature and Reportage*. Pretoria: UNISA Press.

Tucker, Andrew. 2009. *Queer Visibilities: Space, Identity, and Interaction in Cape Town*. Oxford: Wiley-Blackwell.

Tucker, Andrew. 2010. "Shifting Boundaries of Sexual Identities in Cape Town: The Appropriation and Malleability of Gay in Township Spaces." *Urban Forum* 21:107–122.

Tutu, Desmond. 1994. *The Rainbow People of God: The Making of a Peaceful Revolution*. New York: Doubleday.

Valji, Nahla. 2003. "Creating the Nation: The Rise of Violent Xenophobia in the New South Africa." Master's thesis, York University.

van der Vlies, Andrew. 2011. "Zoë Wicomb's Queer Cosmopolitanisms." *Safundi* 12 (3–4): 425–444.

Vena, Jocelyn. 2011. "Beyoncé 'Nailed It' in 'Girls' Video, Choreographer Says." MTV (May 19). www.mtv.com/news/articles/1664223/Beyoncé-run-the-world-girls.jhtml.

Vigneswaran, Darshan. 2007. *Free Movement and the Movement's Forgotten Freedoms: South African Representations of Undocumented Migrants*. Oxford: Refugee Studies Centre.

Wacquant, Loïc. 2001. "Deadly Symbiosis: When Ghetto and Prison Meet and Mesh." *Punishment and Society* 3 (1): 95–133.

Warren, Kenneth. 1993. "Appeals for (Mis)recognition: Theorizing the Diaspora." In *Cultures of United States Imperialism*, edited by Amy Kaplan and Donald E. Pease, 392–406. Durham, NC: Duke University Press.

World Bank. n.d. *Gini Index (Estimate)*. https://data.worldbank.org/indicator/SI.POV.GINI. Accessed September 21, 2019.

Xaba, Thokozani. 2001. "Masculinity and Its Malcontents: The Confrontation between 'Struggle Masculinity' and 'Post-Struggle Masculinity.'" In *Changing Men in Southern Africa*, edited by Robert Morrell, 105–126. Pietermaritzburg: University of kwaZulu Natal Press.

Young, Hershini Bhana. 2006. *Haunting Capital: Memory, Text, and the Black Diasporic Body*. Lebanon, NH: Dartmouth College Press.

Discography

Boom Shaka. 1998a. "Free." Track 4 on *Words of Wisdom*. EMI Records, CDRBL 262.

Boom Shaka. 1998b. "Nkosi Sikelela." Track 1 on *Words of Wisdom*. EMI Records, CDRBL 262.

Boom Shaka. 1999. *Lerato*. Bula Music, CD BULA 052.
Ishmael. 2001. "Roba Letheka." Track 1 on *Roba Letheka*. EMI Records, CD ART 140.
Mafikizolo. 1997. *Mafikizolo*. Kalawa Jazmee Records, CD TEL 3022.
Mafikizolo. 2000. *Gate Crashers*. Kalawa Jazmee Records, CD TEL 3022.
Mafikizolo. 2002. *Sibongile*. Kalawa Jazmee Records, CD TEL 3023.
Mafikizolo. 2003. *Kwela*. Kalawa Jazmee Records, CD TEL 3024.
Mafikizolo. 2005. *Van Toeka Af*. Kalawa Jazmee Records, CD TEL 3057.
Mafikizolo. 2006. *Six Mabone*. Kalawa Jazmee Records, CD RBL 370.
Mafikizolo. 2013. *Reunited*. Kalawa Jazmee Records, digital album. https://itunes.apple.com/za/album/reunited/id635353019.
Mandoza. 1999. *9 II 5 Zola South*. EMI South Africa, CD CCP 20002.
Mandoza. 2000. *Nkalakatha*. EMI South Africa, CD CCP 2012.
Mandoza. 2002. *Tornado*. EMI South Africa, CD CCP 2022.
Mandoza. 2004. *MDZ/Mandoza*. EMI South Africa, CD CCP 2036.
Mandoza. 2013. *Sgantosontso*. Sheer Sound CC, digital album. play.spotify.com/album/277ANcTF7D0FwoPvwJz8Tg.
Mathosa, Lebo. 2004a. "Awudede." Track 2 on *Drama Queen*. EMI Records, CD CCP 2040.
Mathosa, Lebo. 2004b. "Dangerous." Track 10 on *Drama Queen*. EMI Records, CD CCP 2040.
Skeem. 2013. *Waar Was Jy*. https://open.spotify.com/album/05ag4opSVE8eoLno8rmW9q. Accessed September 23, 2019.
Thebe. 2001. "Bula Boot." Track 1 on *Chizboy*. Sony, COL 8157.
Wodumo, Babes. 2016. *Wololo*. https://open.spotify.com/album/orXW55Gl0oKlBFqfXfpv7V. Accessed September 17, 2019.
Zola. 2002. "Ghetto Fabulous." Track 1 on *Yizo Yizo 2*. CCP Record Company, CD YIZO WL 2.

Filmography

Burke, Ed. 2011. *Beyoncé: Year of Four*. Gray Cat Productions, Parkwood Pictures, MTV/VH1 Films.
Hood, Gavin. 2005. *Tsotsi*. Film. Miramax.
The Journey—The Hits from Mafikizolo. 2006. DVD. Kalawa Jazzmee Records. RBM.
Mandoza: Live in Concert at Sun City. 2004. DVD. CCP World.

Videography

"Below the Belt Lebo Mathosa." YouTube. http://www.youtube.com/watch?v=Yw92xDKrUeg. Posted January 2009.
"Below the Belt South Africa—Sexy & Cheeky Adult/Variety/Travel Series Trailer." Underdog Productions. 2013. vimeo.com/74713061.
Beyoncé. "Run the World (Girls)." YouTube. https://www.youtube.com/watch?v=VBmMU_iwe6U. Posted May 2011.
"Desire Marea: Siyakaka a Healing Manifesto." YouTube. https://www.youtube.com/watch?v=kFkgN9PGdMM. Posted September 2017.

"Izikhothane—Part I." YouTube. http://www.youtube.com/watch?v=IWEcV_Ecfl4. Posted May 2012.

"Izikhothane—Part II." YouTube. http://www.youtube.com/watch?v=JbsnhbRM7P4. Posted May 2012.

"Izikhothane—Part III." YouTube. http://www.youtube.com/watch?v=O97BOeLVI5o. Posted May 2012.

Mathosa, Lebo. "Awudede/2Dangerous." YouTube. https://www.youtube.com/watch?v=e-2yt-Wl_uo. Posted April 2007.

Mathosa, Lebo. "I Love Music." YouTube. https://www.youtube.com/watch?v=GdPMC9jootI. Posted May 2008.

"Official Mafikizolo ft. Uhuru Khona." YouTube. https://www.youtube.com/watch?v=yhk52GlkhVA. Posted April 2013.

Rolling With: Mandoza (season 1, episode 1). YouTube. https://www.youtube.com/watch?v=VtiqwnuKGeA. Posted September 2013.

Rolling With: Mandoza (season 1, episode 7). YouTube. https://www.youtube.com/watch?v=le7Ny4vb7QM. Posted October 2013.

Skwatta Kamp. "The Clap Song." Youtube. https://www.youtube.com/watch?v=HL7QWio_iO8. Posted March 2006.

"Two Women Break Down Stereotypes in Pantsula Dancing." YouTube. https://www.youtube.com/watch?v=_LCyaGV3Xoc. Posted May 2017.

Index

Page numbers in italics refer to illustrations

AbbaShante (group), 198
accessibility, 167–68, 181
Africa, 30, 38–39, 68–69
African Americans, 4, 23, 49–50, 77; and hip-hop, 29–30, 37, 49–50, 58, 157–58; musical influence of, 18, 50, 58, 99, 192, 211–13, 228, 237n8; 1980s experiences, 82–83; in South Africa, 32–35, 81–82, 101–2, 147
African Diaspora studies, 11, 29–30
African Methodist Episcopalian Church, 51
African National Congress (ANC), 6–10, 53, 194, 215, 218, 239
Afrikaans language, 33, 217
Afripop! magazine, 208
Afrodiasporic musics, 11, 50–51, 71, 75, 235n5; African American influences on, 18, 50, 58, 99, 192, 211–13, 228, 237n8;

Beyoncé's source material, 37–38; g'qom style, 119, 227, 240; "international," 50–51; jazz, 192–93, 237n6, 241; kwela, 204; marabi, 201, 203–4, 206, 237n6; neotraditional, 198–201; ragamuffin/ragga(muffin), 48; R&B, 71, 198, 237n4; reggae, 72; "Run the World (Girls)" (Beyoncé), 36–42; syncretic cultural combinations, 51; township disco, 2, 22. *See also specific types of music*
Afrodiasporic Space, 5, 19, 25, 27; as analytic concept, 29–32; apartheid and, 71, 82–83; cultural trauma in, 199–94, 202, 221; and food and identity, 166–67; geographies and, 78, 87; global Black cultures, 4, 17–19, 30–31, 38–39, 47–50, 189; Mathosa and, 123–24, 135–36, 142–43, 150, 153–54; misrecognition in, 32–35, 40, 42, 52; popular cultures, 16–18;

Afrodiasporic Space (continued)
  queer rights issues in, 213–14; townships and, 80; tsotsi in, 157, 165–66. *See also* diaspora
"Afro Digital Migration," 50
Afrofuturistic fashion, 214
Afro-Indian identity, 17
Afro-pop, 197, 209
agency, 5, 8, 10, 13; Black women's, 127, 129, 131–34, 137, 146–47; consumption and, 97, 106–7, 112–13; masculine, 156, 171, 186–87
Alexandra, 59, 66
alienation, 159–60
Allen, Jafari, 53
Allen, Lara, 210
alternative practices, 10, 60, 127; "dangerous woman" as, 128–29; futures, 193, 208, 221, 223–32, 234; political vernaculars, 13–14, 224–28, 232, 234; queering of identities, 134–35, 142; Youth Day celebrations, 221, 223
alters, 134, 142
ama kip kip (candied popcorn), 216, 239
anthemic songs: Black, 45–46
anti-apartheid movement, 6, 97, 127, 160–61, 218–21, 232–34
apartheid, 5–10; cultural essentialism of, 192–93; denial of Black projects, 93–94, 97; language enforcement, 217; liberalization of, 20; memory of, 188; pass laws, 21, 62, 88, 204, 207, 241; plantation comparison, 192–93; residual traces of, 113; Sophiatown and, 194–95; women and, 125–26. *See also* white supremacy
archive, 45, 223, 230; performative, 207, 217
Arnfred, Signe, 141–42
aspiration: language of, 16
authenticity, 17, 113, 161–71, 177, 181
autochthony, 47
"Awudede/Dangerous" (Mathosa), 26, 123–24, 138–39, 150–53

Baartman, Sarah (Hottentot Venus), 125
Bad Boy label, 111
"bad girl" persona, 26, 123–24, 127–29
Bailey, Marlon M., 129, 182
Balobedu people, 150–51
Banton, Buju, 48
bare (fool) figure, 166

Barnett, Clive, 147
battles: hip-hop, 39–40; s'khothane, 117–20
Below the Belt (television show), 145–46
Beyoncé, 31, 36–42, *41, 42,* 154
Biko, Steve, 49, 233
binaries, 18, 71–72
bisexuality/"double adaptor," 124, 143–46
Black Economic Empowerment scheme (BEE), 8, 102, 239
Blignaut, Charles, 43, 45
bodies, Black: under colonialism, 159–60; and criminality, 164–65; as dangerous, 88; groups of, 108; and hypersexuality, 123–25, 137; moral panic about "Black township youth," 218–21, 223; policing of, 32–35, 81–82, 106, 147, 164; remastery of suffering, 215–21; sexualized thug persona, 168–70, *169;* white supremacy and, 149. *See also* kwaito bodies
Boikanyo, Refilwe, 117–18
Bond, Patrick, 8
Boom Shaka (kwaito group), 25–26, 125, 143, 198, 230–32, 234, 236n7; "Free," 54; "Nkosi Sikelel' iAfrika," 25, 31, 42–56; *Words of Wisdom,* 55–56. *See also* Mathosa, Lebo (kwaito performer)
Bophuthatswana, 22
Bowes, Greg, 50
Braamfontein, 225
Brah, Avtar, 30
Brickz (kwaito performer), 69
Brookes, Heather, 165–66
Browne, Simone, 81
Buce, Mario (Tofo Tofo), 39
"Bula Boot" (Thebe), 140–41
Burke, Ed, 37
Butch Queens Up in Pumps: Gender, Performance, and Ballroom Culture in Detroit (Bailey), 129
Butler, Judith, 174

Cape Town, 61, 144
Capital (club), 57–60, 64
capitalism: ghetto fabulousness and, 113; global, 3–4, 7–8, 10, 51, 62, 106–7, 171–72, 177; redemptive politics and, 15–16, 171–72. *See also* consumption; neoliberalism
carceral state, 164
Cardi B (Belcalis Almánzar), 226
Caribbean dancing, 69, 141, 150, 157

260 Index

Cheaters (television and radio show), 209
Chicago, 49
Chiskop (group), 156, 167, 168
choreography, 36–42
cis-hetero-topia, 232, 234
citizenship, 13, 46, 65, 147; consumption and, 93, 96, 121; cultural, 178; second-class, 82–83, 88
City Press (newspaper), 38–39, 180–81
Civic Theatre of Johannesburg, 44
"Clap Song" (Skwatta Kamp), 88
class relations, 58, 74–76, 184–86, 192. *See also* middle class, Black
Clegg, Johnny, 149, 240
Cleo (DJ), 69
clothing. *See* sartorial practices
club scene/nightlife, 22–23, 67, 74–75; Capital, 57–60, 64; door policies, 57–59, 78; House of Tandoor, 2–3, 48; Meli's, 85–86; Oh!, 77–78, 98; political dimension of, 88; Rock, 82, 83–85, 183. *See also* partying
colonialism, 5, 11–12, 81
Combs, Sean, 11
commemoration, 189–90
commodification, 16–17, 26, 67, 171; of Sophiatown, 202–3; of tsotsi persona, 156, 176–78, 186–87. *See also* consumption
comrade-tsotsi, 160–61
Congolese dance, 47, 48, 200
Congress of South African Students, 218
Congress of South African Trade Unions, 7
Conquergood, Dwight, 23–24
consumption, 11, 14–16; Black good life and remastery of, 106–7; citizenship and, 93, 96, 121; countering of conventional views of Blackness, 106, 121; freedom and, 25–26, 97–98, 106, 112–13, 120–21; patterns of, 120–21; self-fashioning and, 97–98; sumptuary laws, 25–26, 97, 120. *See also* commodification
contamination, 125–26
conviviality, 60, 85, 189–91, 195, 210
Cooper, Carolyn, 134, 141
Coplan, David, 51, 190, 192
Côte d'Ivoire, 105
counteridentification, 127–28, 153, 159–61
Cox, Aimee Meredith, 144
creativity, Black, 68–69, 87–88
crime, 157–59, 161–65

critical cultural studies, 24
Cuba, 53
cultural formation, 20–22
cultural imperialism, 17–18
cultural labor, 6, 125, 164, 187, 234
cultural production, 24, 109–10, 192, 211
cultural studies, 24, 52
cultural trauma, 190–94, 202, 221

Daily Sun, 136
Dana, Simphiwe, 153
dancehall. *See* Jamaican dancehall
dance moves, 36–38; battles, 39–40, 117–20; Caribbean, 69, 141, 150, 157; circles, 140; Congolese, 47–48, 200; kwassa-kwassa, 48; mixture of, 139–40, 150; ndombolo, 48; pantsula, 37–42, 41, 42, 117; vogueing and waacking, 213–14, 237n8; wining, 48, 138; Zulu, 200. *See also* Jamaican dancehall; movement/mobility
danger, 184–85
"dangerous woman" persona, 26, 123–24, 128–46, 229
death, 220–21
Delaney, Samuel, 67
Democratic Alliance Party, 240
Department of Arts and Culture, 55
desire, 58, 80–83, 103–4, 156, 182–83
Detroit, 182
diaspora: and Blackness, 82; and culture, 18; as embodied, 23; as epistemic practice, 14; and intimacy, 194; and longing, 80, 85, 236–37n1; practice of, 32; as theoretical concept, 30. *See also* Afrodiasporic Space
diaspora studies, 17–18, 29–31, 56
Diawara, Manthia, 16, 106, 178
Diepkloof Extension, 66
"Die Stem" (apartheid anthem), 45
dis battles, 118–19
disco, 2, 22
discourses: alternative political vernaculars, 13–14, 224–28, 232, 234; local-global discursive debate, 17; nationalist, 149–50; public, 5, 113, 202–3; resistance/co-optation (capitulation) binary, 4, 15–17
disidentification, 235n4; critical disappointment, 161; Mathosa and, 26, 123–24, 129, 144–45, 153; tsotsi (thug) persona and, 156, 161, 165, 168, 172, 175–77, 187

Index 261

displacement, 80–81, 193–94, 196, 236–37n1
DJ Cleo, 69
DJ Lynnée Denise, 50
Dlamini, Jacob, 190–91, 195
Dlamini, Nsizwa, 67
DMX, 168
dog: as symbol of hustling masculinity, 165, 216
Dolby, Nadine, 178
domesticity, 173–74, 208–11, 213
domestic workers, 62–64
domination: geographies of, 78
double adaptation/bisexuality, 124, 143–46
drag, 142
Drama Queen (Mathosa), 136
Dream (Mathosa), 133
D-Rex, 216
Drum magazine, 136, 193, 204, 209
Du Bois, W. E. B., 30
Durban, 17, 72, 81, 84, 140, 240

Eaton, Curwyn, 181
Edwards, Brent Hayes, 32
ekasi (township), 113, 165, 240
elites, Black: Black Economic Empowerment scheme (BEE), 8, 102, 239; consumption patterns, 120; cosmopolitanism of, 51; in government, 8, 10, 13, 232; and June 16 celebrations, 189, 218; misreading of kwaito bodies, 27, 189; neoliberal strategies, 7–9; in northern suburbs, 61–62; rehabilitation project of, 156, 171–72, 176
Ellapen, Jordache, 113
Elsdon, Gerry Rantseli (Williams), 124
embodied practices, 11–13, 19, 56, 110–11, 231–34; aural popular histories, 190; of diaspora, 23; intersectional, 51; of musical memory, 190. *See also* bodies, Black; dance moves; performance; sartorial practices
"Emlanjeni" (Mafikizolo), 207
encounter, 56, 60, 67–87, 93–98
English language, 70–74, 81, 99, 217
epistemology, 130
the erotic, 46, 67–68
ethics of knowledge, 130, 235n5
e.tv, 93, 177
excess, 26, 93, 127, 185; as dangerous, 218–21; Mandoza and, 176–77; remastery of, 220–21; self-fashioning and, 98, 111–14, 119, 121. *See also* pleasure
exclusion, 58–60, 156
exuberance: and Black good life, 106; in "I Love Music," 86–87; at kwaito festivals, 108–9; in "Nkosi Sikelel' iAfrika," 46–47; politics of, 26–27, 37, 93, 98, 114, 121; at Youth Day celebrations, 220–21
Eyerman, Ron, 189–90, 201

fabulousness, 111–21, 177
FAKA, 230–32, *231*, 234
"faka," 230
false consciousness, 15, 106
family man, 172–77
Fanon, Frantz, 8
Fassie, Brenda, 123–28, *128*, 135, 143, 153–54, 198, 225–26, 235n3
femininity, 122–54; alternative, 147–48; "bad girl" persona, 26, 123–24, 127–29; cult of, 126–28, 153; "dangerous woman" persona, 26, 128–46; and decency, 136–37; power and, 43, 46, 125, 127, 131, 141–43, 151–52, 154, 225; in Sophiatown, 208, 210; transgressions of, 128, 136–37. *See also* women, Black
feminism, Black, 23–24, 56, 137, 141, 151–52; ratchet, 227–30; Siyakaka, 230–32
Fink, Katharina, 204, 207, 217
Fosse, Bob, 37
4 (Beyoncé), 36–42
freedom: consumption and, 25–26, 97–98, 106, 112–13, 120–21; ethics of, 224; inherited models of, 92; kwaito as expression of, 5, 10–11; neoliberal and neo-Marxist views of, 52; personal, 53–54; remastery of, 12–14, 24–26, 44, 52–55, 68, 93, 123, 132, 154, 232–34; self-fashioning and, 105–6
Freedom Square (Kliptown), 66
"From Jo'burg to Jozi" (Roberts), 60–61
Fugard, Athol, 158

gangs/gangstas, 26, 39, 54, 117, 157–58, 166, 182
Gaston, Frank, 37–38
Gauteng, 22, 80, 114–17. *See also* Johannesburg; Soweto

gender and race: challenges to, 10–12, 18–20, 88, 92, 228–29; conviviality and instability of, 60; in Great Britain, 190; mixture and, 63–64, 67, 78–79. *See also* race

gender and sexuality, 18–19; constructed nature of, 138; dancing and, 40, 140–41; "gangster-sexual," 182; hypersexuality, 123–25; rain queen performance, 150–52; sadomasochism, 139; same-sex desire, 137–38, 142, 236n10; women's nonnormative performance, 43–44. *See also* femininity; masculinity

gender-nonconforming performances, 230–32

geographies: of commodification, 16; of domination, 78; of exclusion, 59–60; of performance, 87; sonic, 92; of space, 25, 60–61, 78

Gershon, Tanja, 75

"ghetto," 111–12

"Ghetto Fabulous" (Zola), 113

ghetto fabulousness, 111–21, 177; hip-hop and, 99–100, 108–9, 111–12, 120

Gilbert, Jeremy, 110

Gilroy, Paul, 60, 85, 190, 194

Gini coefficient, 9, 240

global Black cultures. *See* Afrodiasporic Space

global economy, 3, 7, 20

Gqola, Pumla Dineo, 124, 126–27, 147, 150, 210

g'qom style, 119, 227, 240

Great Britain, 48–49, 190

Gregg, Robert, 30

Group Areas Act, 62, 65

Growth Employment and Redistribution, 7–8

Gucci, Fela (Thato), 230–32, *231*

Guilbault, Jocelyne, 72–73

Guz 2001 (TKZee Family), 51

Hannerz, Ulf, 192

Hansen, Thomas Blom, 17, 79

Harlem Renaissance, 192

Harris, Calvin, 58

Haupt, Adam, 171, 186

heteropatriarchy, 232–33; family man, 172–77; masculinity and, 115, 156, 172–73, 223, 229; Mathosa's challenges to, 123, 131, 153–54, 232; misrecognition and, 32, 35; resistance to through song, 46–47; same-sex desire counters, 137–38; s'khothane queering of, 119; of state, 13, 35, 46–47, 52, 56, 149–50; women and, 126–27, 131, 138, 173

Hillbrow, 59, 65, 124–25

hip-hop: African Americans and, 29–30, 37, 49–50, 58, 157–58; battles, 39–40; ghetto fabulousness and, 99–100, 108–9, 111–12, 120; "keeping it real," 167–68; motswako and, 74; at Turbine Hall, 88; West Coast (Los Angeles–based) "gangsta rap," 157–58

Hirsch, Marianne, 201–2

history, 188–90, 195, 199

HIV/AIDS, 111, 170–71

hola 7, 240

"homespace," 82

Hood, Gavin, 158

hooks, bell, 137

Hope, Donna, 157

house music, 47, 49–51, 58, 69, 77

House of Tandoor, 2–3, 48

Huddleston, Trevor, 194

Hughes, Langston, 32

hypersexuality, 123–25, 137

ikota (sandwich), 166, 240

"illegal immigrants," 33–35, 88

"I Love Music" (Mathosa), 86–87

imagined community, 31

immigration laws, 35

Impey, Angela, 18, 135

indigeneity, 30

"Indoda" video (Mandoza), 177–81, *179*

inequality: masculinity and, 18–19, 173; structural, 8–10, 173, 219

initiation rites, 141–42, 236n9

"In Search of Tommie" (Wicomb), 152

"international" music, 50–51

intersectional analysis, 11–12, 19, 23–24

intraracial concerns, 156–58, 190, 215

Intro (Mathosa), 135

Iqani, Mehita, 97, 98, 113

Ishmael (kwaito performer, Skeem), 1–2, 5

Iton, Richard, 18

Ivies subculture, 117

izikhothane. *See* s'khothane (izikhothane) culture

"Izinja" (Mapaputsi), 216

Jabulani Mall, 94
Jacobs, Jane, 67
Jaji, Tsitsi, 27, 30
Jamaican dancehall: Boom Shaka and, 47–49; and kwaito, 18, 69, 71; Lady Saw and, 134, 141; Mathosa and, 138–39, 150; in "Run the World (Girls)" video, 36; rude boy figure, 157–58, 161; self-fashioning in, 105–6
Jam Alley (teen game show), 235n5
Jamison, Andrew, 189–90, 201
Jay-Z (kwaito performer), 27
jazz, 192–93, 237n6, 241
Jazz (ragga-rapper), 138–39, 150, 152, 236n8
Jeppestown, 66
Jim Comes to Jo'burg (film), 210
Johannesburg, 9, 21, 61–62, 63; apartheid in, 89–91; CBD, 64–65, 80; Congolese migrants, 48; gentrification, 66; inner-city area, 64–65; perspectives on, 60–61; redevelopment, 64–67, 88–90; remastery of, 87–91. *See also specific neighborhoods*
Johannesburg Development Agency, 65–66
Johnson, E. Patrick, 98, 131, 143
Jones, Omi Osun Joni, 23, 24
Jules-Rosette, Bennetta, 51
June 16 parties. *See* Youth Day celebrations (June 16)
JUNGLEPUSSY (Shayna McHayle), 226

Kabir, Ananya, 189, 190, 201–2
"Kaffir" (Mafokate), 10–11
KasieKulture blog, 137
Kau, Irene, 194
Kelley, Robin D. G., 88
Kenya, 13
Keyes, Cheryl, 157
Kgosinkwe, Theo (kwaito performer), 198, 199, 202, 205, 205–7, 209, 211
Kharsany, Safeeyah, 163–64
Khelobedu language, 150
"Khona" (Mafikizolo), 213–15, 223
Khumalo, Kelly, 181
kleva identity, 157, 165–66, 240
Kliptown, 66
kofifi parties, 201, 237n7
Kruger, Loren, 191
kwaito bodies, 4, 19, 24–27; as concept, 12–15, 31; encounter and, 60, 68; fans, 5, 10, 13–14, 24–25, 56–57, 131, 142, 153, 168–69, 189; illegibility of, 35, 47; and June 16 suffering, 189; meanings of, 31, 47, 52–53, 56, 68; memory and, 189, 195, 223; nation and, 55–56; tradition and, 200; trauma and, 189, 221. *See also* remastery
kwaito festivals, 98–111. *See also* partying; Youth Day celebrations (June 16)
Kwaito's Promise: Music and the Aesthetics of Freedom in South Africa (Steingo), 19–20
kwassa-kwassa dance, 48
Kwela (album; Mafikizolo), 195, 203, 204, 207
"Kwela" (song; Mafikizolo), 207, 210
kwela music, 204

labor: confines of, 88; cultural, 6, 125, 164, 187, 234; gendered, 115–16, 125; migrant, 40, 48, 51, 114–17, 206–7, 236–37n1; in music industry, 56, 144; of performance, 129–30, 135–36, 153–54, 162, 214; of tsotsi, 163–64, 171
Lady Saw (Marion Hall, dancehall queen), 134, 135, 141
Laka, Don, 125
languages: mixture of, 69–74; slangs and argots, 158, 166. *See also specific languages*
Leclerc-Madlala, Suzanne, 149
legibility/illegibility, 35, 47, 182
leisure spaces, 24–25, 60–64, 67, 72, 87–88, 97–98, 158. *See also* space
Lemonade (Beyoncé), 37
lesbian identity, 142–45, 152, 236n12
liberation movements, 6, 27, 51–54, 160, 189, 232–34
Lil' Kim (Kimberly Jones, hip-hop artist), 134, 135, 142
liminal spaces, 65
Lindela facility (South Africa), 33–34
Lindwa, Buhle, 228–29
Lipsitz, George, 106
Lonmin (British-based mine owners), 9–10
Lorde, Audre, 46, 55, 233–34
"Lotto" (Mafikizolo), 198, 200, 213
Loxion Kulca (clothing company), 99
Lynnée Denise (DJ), 50

Mabandu, Percy, 116
Macharia, Keguro, 12, 224
Madikezela-Mandela, Winnie, 220–21

Madingoane, Tebogo, 198, *199*, 202, 204–5, 205, 209
Madondo, Bongani, 127, 154
Mafikeng (capital of Northwest Province), 74
Mafikizolo (group), 26–27, 189–90, 195–223; "Emlanjeni," 207, 209; *Gate Crashers*, 198, *199*; "Khona," 213–15; *Kwela*, 195, 203, 204, 207, 209; "Kwela," 207, 210; "Lotto," 198, 200, 213; "Majika," 198–201, 208–9, 214; "Makhwapheni," 209; "Marabi," 203, 206, 210; "Mas'thokoze," 209; "Ndihamba Nawe," 205–10; *Reunited*, 211–12, *212*; "Sebenza," 207–8; *Sibongile*, 195, 201–4, 202, 207; "Sibongile," 202; *Six Mabone*, 195, 203, 206, 210; "Udakwa Njalo," 209; *Van Toeka Af*, 195, 203, 204, 207–8
Mafokate, Arthur, 10–11, 50–51
Magubane, Zine, 18
Mahlatini, 48, 135
Mahlerbe, Petrus, 220
Mahmood, Saba, 110
Mail and Guardian, 45
"Majika" (Mafikizolo), 198–201, 208–9, 214
majita (guys), 84, 240
Makgoba, Malegapuru, 148–54
Makhulu, Anne-Maria, 61
makwerekwere (derogatory term for immigrants), 33
malls, 61, 93–98, *95*, 100
Malope, Rebecca, 56
Mandela, Nelson, 7, 20, 49, 93–94, 188, 213; and Boom Shaka, 230–31
Mandoza (kwaito performer), 26, 155–87, 169, 220, 233, 236n1; *Godoba*, 160; "Indoda" video, 177–81, *179*; *9 II 5 Zola South*, 156, 159, 162; *Nkalakatha*, 168; *Sgantsontso*, 177; "Skelegeqe," 175–77; *Tornado*, 165. See also tsotsi (thug) persona
Manuel, Peter, 141
mapantsula, 117
Mapaputsi (kwaito performer), 216
Maponya, Richard, 62, 93–94
Maponya Mall (Soweto), 26, 93–98, *95*
"Marabi" (Mafikizolo), 203, 206, 210
marabi music, 201, 203–4, 206, 237n6
Marea, Desire (Buyani Duma), 230–32, *231*
Marikana mine workers' strike, 9–10, 164, 234
Marley, Bob, 2

marriage, 172–75; lobola (money paid by groom's family), 172, 240; same-sex, 173; weddings, 199–200, 205–6, 208–9
Marumo, K., 74
masculinity, 4, 26, 155–87; denaturalizing, 138; dog as symbol of, 165, 216; female, 216; inequality and, 18–19, 173; normative, 18; pantsula-style dancing and, 40; parading practices, 106; self-fashioning and, 114–15
Masekela, Hugh, 192–93, 207
Masemola, Thami, 131–32
Masondo, Amos, 93–94, 194–95
mastering, 12–14
Masters at Work, 198
master-servant binary, 71
Masuka, Dorothy, 207–8, 241
materiality, 14, 26, 31, 61, 63–64, 107, 117, 176. *See also* capitalism; consumption
Mathosa, Lebo (kwaito performer), 26, 42–54, *43*, *44*, 86–87, 122–54, *151*, 225–26, 229; "Awudede/Dangerous," 26, 123–24, 138–39, 150–53; *Drama Queen*, 136; *Dream*, 133; *Intro*, 135. *See also* Boom Shaka
Matsho, Thabo, 194, 237n3
Mattera, Don, 192–94
Mazibuko, Lindiwe, 52, 54
Mazwai, Thandiswa, 153
Mbembe, Achille, 66, 104
Mbulu, Letta, 207
McAdoo, Orpheus, 51
McEachern, Montinique, 228
Mchunu, Harry, 171
McKittrick, Katherine, 78–79
Mdlongwa, Oscar, 198
mediascapes, 186
"Meet Me at the River" (Mbulu), 207
melancholia and melancholic conviviality, 80, 85, 189–91, 195, 210
Meli's (club), 85–86
Melville, 75–77
memory, 27, 220–23; collective, 190; disremembering of house genealogy, 49; musical, 189–90, 195, 201–2; nostalgia, 80, 85, 190–91, 195; post-memory, 201–2, 207; queering of, 188, 221; re-memory, 51; selective, 192; utopian, 188, 191–92. *See also* past
Metro FM, 76

Index 265

"Mexican Breakfast" routine (Fosse), 37
Mgcina, Sophie, 207
Mhlambi, Thokozani, 17
middle class, Black, 55, 62–63, 66, 80, 90–91; assimilation of, 233–34; compensatory consumption, 120; cosmopolitanism, 58; English skills, 165; family man, 172–77; ghetto fabulous opposition to, 112–13; in Sophiatown, 191. *See also* class relations
migrant labor. *See* labor
militarism, 126
Miller, Monica L., 117, 119
Minaj, Nicki, 134
misrecognition, 32–35, 40, 81–82, 102; "Nkosi Sikelel' iAfrika" and, 42, 47, 52
mixed methods, 23–25
mixed-race spatiality, 58, 64, 189–91, 215
mixture: of dance moves, 139–40, 150; intermixtures, 71–72, 84–85, 90–91; of languages, 69–74; motswako, 60, 67, 69, 72–75, 87, 215; nationality and, 70; as process of dispossession, 71; racial, 69, 71
"Mmangwane" (Mgcina), 207
Mnisi, Jabulani, 120–21
Modisane, Bloke, 194
Modjadji. *See* rain queen of the Balobedu
Mofokeng, Mandla, 188
Morafe (hip-hop group), 74
moralist frameworks, 16
moral panic, 218–19
Morgan, Ruth, 152–53
motswako ("mixture"), 60, 67, 69, 72–75, 87, 215
Motswako (television show), 73–74
Mottiar, Shauna, 8
movement/mobility, 4, 11, 31, 70–72; Black bodies and, 81–82, 98; class performances, 75; encounter and, 79–87; in Johannesburg, 62–67. *See also* dance moves
Moynihan Report, 173
Mozambique, 38–40, 141
Mphithi, Luyolo, 63–64
Muholi, Zanele, 154
Muñoz, José, 123, 129, 130, 153, 235n4
musical memory and recall, 189–90, 195, 201–2
musical production, 21–22, 50
music industry, 56, 125, 135–36, 143–44

national anthems, 45. *See also* "Nkosi Sikelel' iAfrika"
nationalism, 56, 148–54
National Party, 6, 8, 215, 218
native nostalgia, 190–91, 195
Nciza, Nhlanhla (née Mafu), 197–98, 199, 202, 203–9, 205, 211
Ndebele, Njabulo, 233
"Ndihamba Nawe" (Mafikizolo), 205–10
ndombolo dance, 48
The Negro (Du Bois), 30
neoliberalism, 3, 5, 7–10, 15, 20–21, 55, 88, 111. *See also* capitalism
neotraditional music, 198–201
New York, 111–12
Ngcaweni, Wandile, 79
Ngidi, Sandile, 46
Nguni languages, 74
Nguyen, Vinh-Kim, 105
Nhlengethwa, Theo (kwaito performer), 42, 43, 143, 216–17
Niaah, Sonjah Stanley, 18, 47
nightlife. *See* club scene/nightlife; kwaito festivals
9 II 5 Zola South (Mandoza), 156, 159, 162
Nixon, Rob, 191
Nkabinde, Nkunzi, 152–53
Nkosi, Lewis, 114–15, 118, 210
"Nkosi Sikelel' iAfrika," 25, 31, 42–56
"Nolishwa" (Masuka), 208
Northwest Province, 74
Norwood, 85, 241
nostalgia, 80, 85, 190–91, 195, 236–37n1
Noyes, John, 139
Number 10 (film), 165
Nuttall, Sarah, 16, 99–100
Nxedlana, Jamal, 115–18
Nxumalo, S'bu (the General), 14

Oh! (club), 77–78, 98
Oshun (Yoruba fertility figure), 141
Oswin, Natalie, 16
Otherness, 88
O'Toole, Kit, 12

pantsula dance and culture, 37–42, 41, 42, 117
parading, 106

partying, 60, 93, 97–98, 103–5, 110–11.
  See also club scene/nightlife; kwaito festivals
passivity, 126
pass laws, 21, 62, 88, 204, 207, 241
past, 5, 10–11, 232–34; Africa framed in, 30, 38–39; Mafikizolo's performance of, 195–207, 211, 214; queering, 26–27, 152; remastery of, 188–91, 195, 197. *See also* memory
Patra (dancehall queen), 47
Pearson, Ewan, 110
People of the South (television program), 132, 165, 241
performance: carceral space and, 164; of death, 220–21; as epistemology, 130; ethnography, 23–24, 67–68; of everyday life, 24; gender and, 155–56, 174, 230–32; geography, 87; of interracial possibility, 156–57; labor of, 129–30, 135–36, 153–54, 162, 214; linguistic, 33; of memory, 189; misrecognition and, 32–35, 40, 52; of nation, 46–47; power of, 125, 131, 134–35, 139, 182; practices, 2–3, 10–11, 24–26, 47, 87, 92, 114, 123, 129, 224; queer women's, 216; reality equated with, 131–33; same-sex desire and, 137–38, 142; of township, 76–77; tsotsi thug authenticity, 161–71, 177. *See also* s'khothane (izikhothane) culture
performative archive, 207, 217
Peterson, Bhekizizwe, 15, 16, 18–19, 46–47, 165
Phatstoki (Gonste More), 225, 226
Pierre, Jemima, 30
Pietilä, Tuulukki, 16, 17–18, 31
Pillay, Suren, 161
Pimville, Soweto, 93, 180–81
Pitch Black Afro (group), 74
pleasure, 11, 25–26, 68, 93; Black women's, 127; as end in itself, 110; expansion of, 43, 46, 55; kwaito as dance music, 109; as political, 20, 111. *See also* excess
Pokwana, Vukile, 43
policing: of Black bodies, 32–35, 81–82, 106, 147, 164; of gender, 231; internal, 233–34; of kwaito festivals, 98, surveillance and, 81
political economy, 5, 10, 12, 17, 20
political vernaculars, 13–14, 224–28, 232, 234
"Pon the Floor" (song), 36

Posel, Deborah, 97, 98
possibility, 25–27; geographies and, 78–79; intraracial and intermixtures, 71–72, 156–57; limited, 10; Jozi and, 59–60, 64–65, 90; Maponya Mall and, 94; redemption and, 172; in Yeoville, 3, 64
post-memory, 201–2, 207
poverty: coordination of social services, 7; Mandoza's narrative and, 156, 162; refusal to be limited by, 16, 97, 111, 117, 121; in rural areas, 21; in townships, 66
power: colonial, 160; feminine, 43, 46, 125, 127, 131, 141–43, 151–52, 154, 225; hierarchies of, 23, 79; kwaito bodies and, 53–54, 68, 90, 105–6; misrecognition and, 32; of performance, 125, 131, 134–35, 139, 182; political, 6, 10–12, 109, 112, 125; structures of, 6, 217, 233; of tsotsi affect, 182, 185
privatization, 8, 21, 66–67
public discourses, 5, 113, 202–3
public sphere, 12, 55, 124–26, 230
pum-pum shorts, 47
pussy parties, 225–27, 226, 229

"quare studies," 143
queering: alters as strategy of, 134–35; of binaries, 18, 71–72; of consumption, 26; of masculinity, 156, 182–86; of Mathosa, 142; of memory, 188, 221; of nationalism, 148–54; of past, 26–27, 152; post-apartheid nationalism and, 148–54; rainmaking trope, 152–53; of sexual interaction, 105; of tsotsi (thug) persona, 168, 182–87; of white supremacy, 71–72
queer space/queer-friendly space, 2–3, 77, 84, 216–17, 220
queer theory, Black, 23–24, 56

race: and consumption, 97; and hierarchization, 71–73; and inequality, 157; performance of interracial possibility, 156–57; and mixture, 69, 71; and spatiality, 58, 64, 189–91, 215. *See also* gender and race
Radio Bop, 22
Radio Metro, 22
ragamuffin/ragga(muffin), 48
ragga chanting/toasting, 47, 48–49, 69
"Rainbow Nation," 7–8

Index 267

rain queen of the Balobedu, 150–53
Ramaphosa, Cyril, 10, 234
R&B, 71, 198, 237n4
Rand Show, 130, 142
Ranks, Shabba, 48
rape, 132, 147, 236n6, 238n13
Rastafarianism, 72
ratchetness, 228–30
Rathebe, Dolly, 210
rave music, 77
realness, 182
reconfiguration, 188, 195, 232; of Africanness, 18; of consumption, 107; of dance styles and music, 37, 40, 48; of gender, 78, 140–41; of Johannesburg, 61, 66, 92; of masculinity, 156–57, 164, 181; of social relations, 75
Reconstruction and Development Programme (RDP), 7
Reddy, Gayatri, 143, 144
redemption/rehabilitation, 15–16, 168; Black men and, 171–72; commodification of, 178; criminal justice system and, 158–59, 162; family man and, 173; of masculinity, 156–60; musical recall and, 190; "prodigal son," 180–81; of tsotsi (thug) persona, 26, 39, 156–60, 171–82
Redmond, Shana, 45–46
Reefenhausen, Baroness Coral von, 145–46
Refashioning Futures: Criticism after Postcoloniality (Scott), 159
reggae music, 72
remaking: of boundaries, 23, 60; of city spaces, 66–67, 79, 87; of racialized colonialism, 12; of respectability, 174; of subjectivity, 5–6
remastery, 19, 224; of American jazz, 192; of consumption, 106–7, 111–14, 121; of excess, 220–21; of freedom, 12–14, 24–26, 44, 52–55, 68, 93, 123, 132, 154, 232–34; of Johannesburg, 87–91; of masculinity, 156, 176, 182–87; micro-practices of, 60; multiple, 186; of past, 188–91, 195, 197; pussy parties, 225–27, 226, 229; refusal of recuperation, 229–30; Siyakaka feminism, 230–32; of suffering Black body, 215–21
re-memory, 51
resistance, 109–12, 127–29, 217–18, 220; counteridentification, 159–61; kwaito and, 4–5, 15; ratchetness as, 228

resistance/co-optation (capitulation) binary, 4, 15–17, 129
respectability, 55, 112–13, 148, 156; anti-respectability/disrespectability, 228; Mafikizolo and, 208–10; Mandoza and, 174, 176–77
retail space, 93
Reunited (Mafikizolo), 211–12, 212
Ringo (pop artist), 198
"Roba Letheka" (Ishmael), 140
Roberts, Adam, 60–61
Rock (club), 82, 83–85, 183
Rogerson, Christian M., 204
Rolling With (reality show), 177, 181
Rosebank, 57, 62–63, 74–75, 100
Rosie Parade (Colleen Balchin), 225–26
rude boy figure, 106, 157–58, 161
"Run the World (Girls)" (Beyoncé), 31, 36–42
"runway," 108, 118–19, 216
rural (homeland) areas, 20–21

sadomasochism, 139
same-sex relationships, 137–38, 142, 173, 236n10
Samuelson, Meg, 150, 208, 210
Sandton/Sandton City, 61–62, 64, 94, 177
sangomas, 152–53
Santoga, Enoch, 51
sartorial practices: Afrodiasporic styles, 40–42, 43, 44, 47–48, 99; competitive dressing, 117; imported brands, 101; at kwaito festivals, 99–101, 102; pantsula dancers, 39–40; in Sophiatown, 203–4; tailors and designers, 100; traditional, 199–200; Y generation, 99–100. *See also* self-fashioning
S'camto (urban vernacular slang), 74, 130, 166, 186
scholarly imaginary, 14–20
Scott, David, 105–6, 159, 161
Sedumedi, Itumeleng, 136
seeing, 166
Seete, Thembi (kwaito performer), 42, 43, 44, 52, 54, 125, 132, 148
self, 121, 132, 134–35
self-care, 129
self-fashioning, 67, 77–78, 93; alcohol and, 103; competitive dressing and, 117;

268 Index

consumption and, 97–98; excess and, 98, 111–14, 119, 121; freedom and, 105–6; ghetto fabulousness as, 112–13; kwaito festivals and, 98–111; "organizing," 104–5; social history of, 114–17; tsotsi persona and, 160–61; *ukukhothana* aesthetic practices, 113–20, *115*, *116*. *See also* sartorial practices
self-theorizing, 23
Setswana language, 74
sexuality. *See* gender and sexuality
Shabazz, Rashad, 164
Shakur, Tupac, 99
shapeshifting, 144
shebeens, 158, 183–84, 201
Sibongile (Mafikizolo), 195, 201–4, *202*, 207
Simone, AbdouMaliq, 87
Singer, Linda, 111
Six Mabone (car), 206
Six Mabone (Mafikizolo), 195, 203, 206, 210
Siyakaka feminism, 230–32, 234
Skeem (kwaito group), 1–2, 5
"Skelegeqe" (Mandoza), 175–77
skelegeqe persona, 166, 175–77
s'khothane (izikhothane) culture, 25, 93, 118–21, 218
Skwatta Kamp (group), 74, 88
slavery, 72, 81, 177, 190
Smith, Andrea, 149
Smith, Danyell, 112
Snorton, C. Riley, 134–35
social capital, 100–105
social hierarchies, 92–93
Sokhela, Junior (kwaito performer), 42, 43, 47, 48–49, 54
solidarity, 35, 45
sonic texture, 49–51, 119–20, 199, 208–9, 237n6
Sophiatown, 189–95, *196*, 201–11, 214–15, 217; destruction of, 26, 189, 223, 237nn2–3; gangs in, 117; Mafikizolo and, 195, 201–9; Tsotsitaal language, 204, 240
Sotho language, 69–70
Sotho-Tswana language, 73–74
South Africa: criminal justice system, 158–59; economics, 6–8; first democratic elections, 4, 7, 20; "new," 3, 25, 53, 234; policing, 32–35, 81; post-apartheid, 8–12, 17, 20–21, 65, 109–10, 229, 232; public culture, 11, 52, 74, 127, 241; queer people in, 213–14; sociopolitical context, 20–22; structural inequality, 8–10, 173, 219; urbanization, 20–22; as "white" country, 33; white men in, 148–54. *See also specific locations*
South African Broadcasting Corporation, 22, 73
South African Communist Party, 7
South African Music Awards (SAMAs), 11, 42–45, 55, 124
South African Police Service, 9–10
Sowearto (boutique store), 100
Soweto, 25, 59, 76, 82–84; Kliptown, 66; Maponya Mall, 93–98, *95*; Zola neighborhood, 161–62. *See also* Gauteng; Johannesburg
space, 75–79; alternative geographies of, 25, 60–61, 78; bodily experience of, 68; conceptual and ideological, 30; of exclusion, 59–60; meanings of, 78–79; Otherness and, 88; remastering, 14, 25, 53, 68, 87–91, 108. *See also* Afrodiasporic Space; leisure spaces; queer space/queer-friendly space
spatial authenticity, 161–62, 166, 181
state: Black women and, 144; co-optation of Black anthemic song, 46; diaspora and, 18, 31–32; and patriarchy, 13, 35, 46–47, 52, 56, 149–50, 174; political vernaculars and, 13–14; violence of, 6, 9–10, 14, 46, 146
Steingo, Gavin, 15, 19–20, 49
Stephens, Simon, 132, 139, 147
Stoned Cherrie (fashion house), 201, 204, 206
Story, Kaila, 32
street parties, 215–16
students, 75–76, 103, 217–18
subjectivity, 14, 30, 123, 129, 134–35, 166. *See also* self; self-fashioning
sumptuary laws, 25–26, 97, 120
Sun City, 162–63, 167
surveillance, 32–35, 81, 106, 144, 164
"Sweety My Baby" (Brickz), 69
Swenka culture, 114–17

Tabane, Rapule, 162
Tambo, Dali, 135
Taylor, Diana, 45
Thebe (kwaito performer), 140–41
Themba's Modern Misses, 208

Thomas, Anthony, 49–50
Thomas, Deborah, 106
thug figure. *See* tsotsi (thug) persona
Times Square (New York), 67
Times Square Red, Times Square Blue (Delaney), 67
TK (R&B star), 176
TKZee Family (group), 51
toasting lyricism, 48
Tofo Tofo (pantsula dance troupe), 37–42, 41, 42
township spaces, 5–6, 21–22, 66–67; as Afrodiasporic Space of resolution, 80; and apartheid, 84–85, 88, 90; *ekasi* as term for, 113, 165, 240; moral panic about "Black township youth," 218–21, 223; performance of, 76–77. *See also* Johannesburg; Sophiatown; Soweto
tradition, 2, 88, 198–201, 214, 220–21
trauma: cultural, 190–94, 202, 221
Trompies (pantsula dance troupe), 188
True Love magazine, 173–74, 209
Truth and Reconciliation Commission, 223
Tshabalala, Mpho (Mandoza's wife), 174–75, 181
Tsotsi (film), 158
tsotsi (thug) persona, 39, 54, 156, 216; and authenticity, 161–71, 177; commodification of, 156, 176–78, 186–87; queering of, 168, 182–87; rehabilitation of, 26, 39, 156–60, 171–82. *See also* Mandoza
Tsotsitaal language (Sophiatown), 204, 240
Tucker, Andrew, 117, 144
Tuks (hip-hop performer), 74
Turbine Hall, 88, 89
Tutu, Desmond, 7
Twala, Chicco, 198
Tzaneen, 152

Uhuru (group), 214
ukukhothana (showmanship) aesthetic practices, 113–20, 115, 116
Umlazi (township), 81
United States, 149, 157, 173, 182
University of Witwatersrand, 80
Unk (hip-hop artist), 37
Urban Areas Act, 191
urbanism, Black, 87–88

urbanization, 20–22
utopian memory, 188, 191–92

van der Vlies, Andrew, 152–53
vernaculars: political, 13–14, 224–28, 232, 234
Vibe magazine, 112
Vintage Cru (dance team), 213–14
violence, 67, 110–11; anti-apartheid movement and, 6; Marikana mine workers' strike, 9–10; sexual, 146–48; Sharpeville massacre (1960), 9; of state, 6, 9–10, 14, 46, 146
Virginia Jubilee Singers, 51
vogueing (dance style), 213–14, 237n8

waacking (dance style), 213–14, 237n8
"Waar Was Jy?" (Skeem), 1–2
Wacquant, Loïc, 164
"Walk It Out" (Unk), 37
Warona, Oscar, 117
Warren, Kenneth, 32
weddings. *See* marriage
White, Tim, 50
white supremacy, 35, 71–72, 85, 148–54, 190, 233–34. *See also* apartheid
Wicomb, Zoe, 152
wining (dance style), 48, 138
Wodumo, Babes, 124, 153, 154, 227, 227–30, 234
"Wololo" (Babes Wodumo), 227
woman: figure of, 207–8
women, Black, 18–19, 122–23; and anti-apartheid movement, 127; bodies of, 146–48; and disidentification, 26, 123–24, 129, 135; DJs, 225–27; and double adaption, 143–44; as kwaito performers, 26, 47, 198; marriage and, 172–75; miniskirt issue, 136; and nationalist discourse, 149–50; and party provision, 104; and passivity, 126; ratchetness, 228–30; retrospectives of Mathosa, 147–48; s'khothane culture and, 119; subjectification of, 129; as threat to social order, 125–26. *See also* femininity; *specific women*
Words of Wisdom (Boom Shaka), 55–56
working-class subject, 159–60
"Wrath of the Dethroned White Males" (Makgoba), 148–54

Xaba, Thokozani, 160
Xhosa language, 33, 70, 74
Xuma family home (Sophiatown), 194

Y generation, 25, 27, 31, 54, 106; clothing, 99–101; and kwaito, 189; and Mathosa, 154; women of, 140. *See also* Youth Day celebrations (June 16)
Y Magazine, 14
Yeoville, 1–2, 48, 59, 65
YFM radio, 14, 54, 131, 137
Yizo (television series), 147
"Yizo Yizo" (Fassie), 235n3

Young, Hershini Bhana, 30
Youth Day celebrations (June 16), 26–27, 131, 189, 215–23, 222, 238n12
YouTube, 37–38

Zola (kwaito artist), 14, 113, 156, 181; on Mandoza, 167, 172–73
Zola (Soweto neighborhood), 161–62
Zola 7, 240
Zone Mall (Rosebank), 100
Zulu dances, 200
Zulu language, 33, 70, 74, 81, 150
Zuma, Jacob, 155, 219, 238n13

www.ingramcontent.com/pod-product-compliance
Lightning Source LLC
Chambersburg PA
CBHW070755230426
43665CB00017B/2375